A STUDY OF THE HISTORY AND DEVELOPMENT OF ANIMATION IN AFGHANISTAN

Author: Mursal Azizi
Supervisor: Shanthi Mathai
Proofreader: S Coleman
Art pieces including cover page: Afghanculture and Afghanculturemuseum
Design: Mursal Azizi
First Published: 2024
10 9 8 7 6 5 4 3 2 1
ISBN 978-0-9756266-2-7
Printed & Bound in Australia
Published by: RoseyRavelston Books

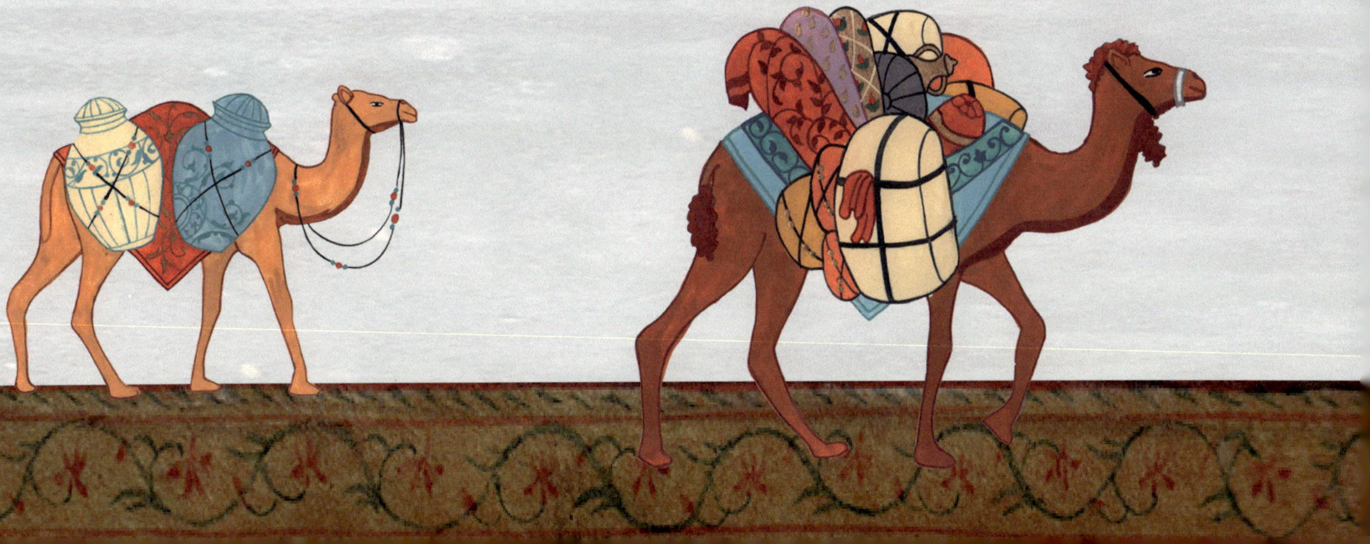

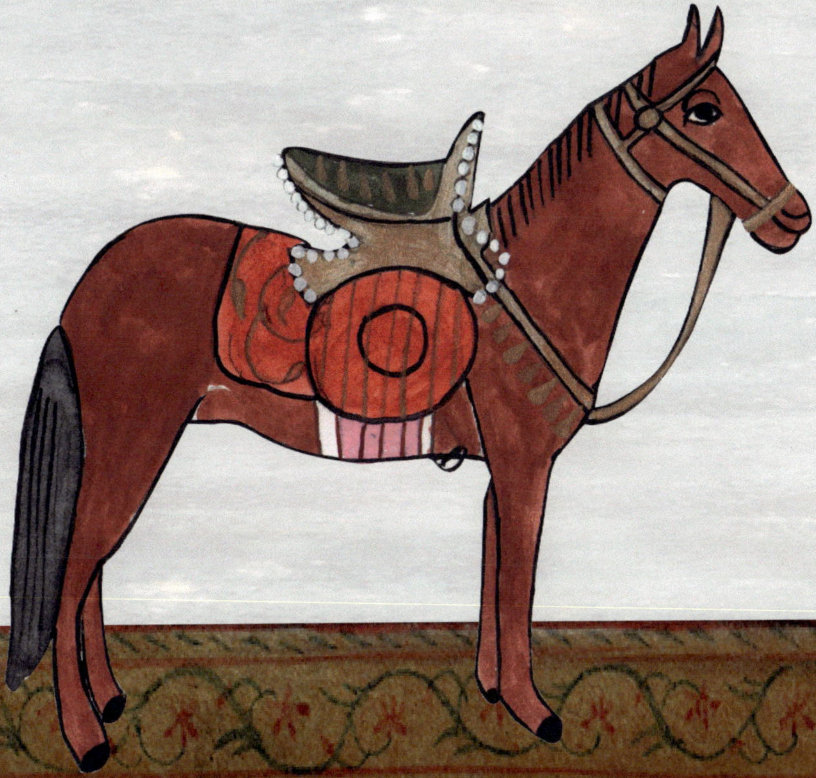

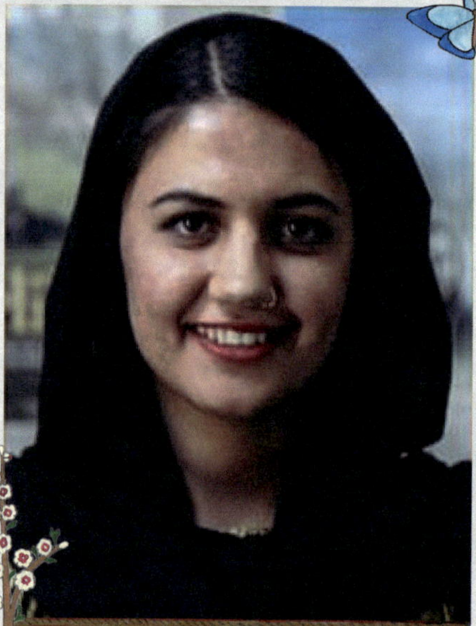

Mursal Azizi, The Author

About the author

Mursal Azizi is an Afghan artist, graphic designer, and digital storyteller. She was born in Kabul, Afghanistan. She completed her Bachelor's degree in Graphic Design from the Fine Arts Faculty of Kabul University in 2015.

She previously worked as graphic designer with CARD-F, a UK based organisation and as an illustration specialist with a project called "Livelihoods Improvement in Tajik-Afghan Cross-Border Areas", a project of the United Nations Development Programme (UNDP).

After completing her Master's degree in Digital Animation from the Sacred Heart School of Communication at Mahatma Gandhi University of India in 2020, she returned back to Kabul and worked as a contracted animator with the US Institute for Peace (USIP). She created two short animations for this project, "Solidarity" and "Dream of Peace", to promote social peace building throughout Afghanistan. She also created a documentary in India on women's empowerment entitled "Chinnamma Story of a Woman Hidden Under Water". The documentary received the 'Best Editing' Award from Herat International Women's Film Festival" (HIWFF) and the 'Best First Film' Certificate from Lajaward National film festival in Afghanistan.

Mursal recently migrated to Australia where she has worked with Animal Logic, one of Australia's leading animation studios, as a digital artist. In 2023, on the second anniversary of the fall of Kabul, Mursal presented an exhibition entitled "Displaced Wishes" at the College of Law, NSW. Here she shared her story, art works and animations in order to fundraise to support several Afghan women in Pakistan.

Mursal is passionate about merging her skills to forge a purpose-driven career that empowers artists and women in Afghanistan and Australia. She has worked as a community volunteer since the fall of Kabul to help Afghan refugees and asylum seekers, mainly women and children.

Table of Contents

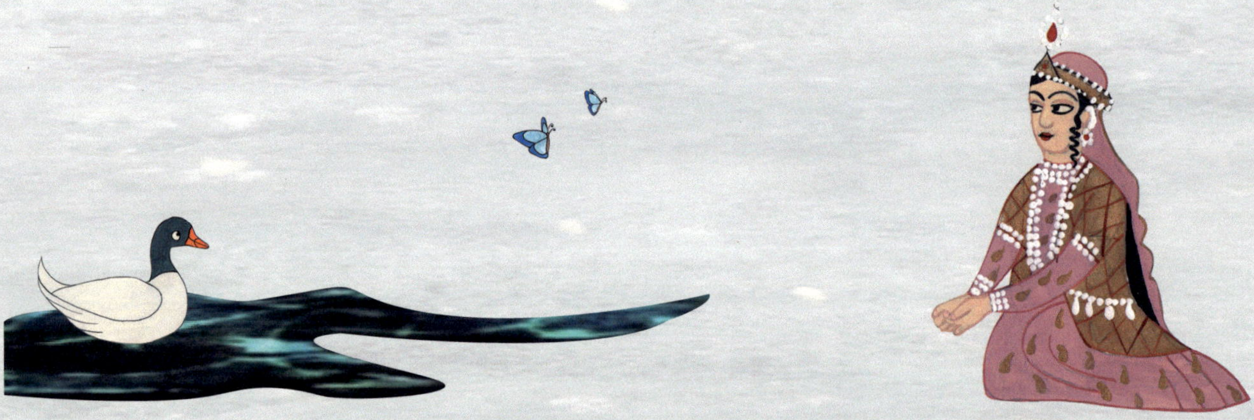

Preface

This book focuses on the history and progress of animation in Afghanistan. The idea for this book goes back to 2018 when I was pursuing my Master's degree in Digital Animation in India. After reading about the history and progress of animation in different countries, I was fascinated by the idea of researching animation in Afghanistan and publishing this book. Until now, there is no written history of "Afghan Animation".

In order to accomplish this goal, I left India for Afghanistan in 2019 to collect the required data personally from different places in Kabul such as the Ministry of Information and Culture of the Afghan Government, Afghan Film Organization, National Radio Television of Afghanistan, Kabul University Library, Fine Arts Faculty, Film and Advertisement Productions, Institut Francais d'Afghanistan IFA and the Academy of Art.

The lack of secondary data such as written documents and online sources was a challenge to achieving my goal. Instead, I interviewed 16 individuals involved in this area of research in Afghanistan. They included different government officials, Afghan film-festival holders, production directors and individual animators.

The names and details of the interviewees are as follows:

- Ms. Hasina Safi: Acting Minister of the Ministry of Information & Culture.
- Ms. Sahra Karimi: Director of Afghan Film Organization.
- Mr. Hezbullah Sultany: Head of Animation Department of Afghan Film Organization.
- Mr. Abdul Hashem Seqiqi: General Manager of National Radio and Television of Afghanistan Archive.
- Mr. Samiullah Nabipur: Dean of Fine Arts Faculty of Kabul University.

- Mr. Habibullah Jahanyaar: General Director of Central Library in Kabul University.
- Mr. Mohsen Hossaini: The first animator following the Taliban regime in 2001 and the first animation instructor in Afghanistan.
- Miss Lajaward Haqiqi: The first female Afghan animator.
- Mr. Zubair Kakar: Creator and performer of puppet animation. He has created Afghan puppets for TV shows.
- Sara Barack: A female Afghan artist who has a Bachelor Degree in animation.
- Mr. Mohammad Ali Kiyan: Director of Mehrgan International Film Festival.
- Mr. Rashed Lovaan: Roya Film House staff.
- Mr. Mohammad Nasir Hashimi: Director of Kapila motion pictures.
- Mr. Mujtaba Karimi: Production Director of Zawia Media.
- Mr. Hamed Elmy: Executive Director and founder of FKH Media.

In addition to interviews, I also conducted some archival studies, including from the Afghan-Film Archive and Archive of National Radio Television of Afghanistan (RTA,) and reviewed many festival brochures, library and online sources.

Currently, the term "Afghan Animation" is perhaps unknown both to the people in Afghanistan and the world outside. I hope that through this research, this book will be a starting point to create a name for "Afghan Animation". I believe this book will help the readers to learn about animation in Afghanistan and the challenges and opportunities in this area. I hope it I will also pave the way for those who want to do further research on this topic.

The objectives of this book are:

- To understand and document the importance given to animation in the history of art in Afghanistan.
- To trace the progress of animation after 2001 in Afghanistan (including animators, movies, studies, television, festivals and awards).
- To identify types of animation used in Afghanistan (including traditional, 2D and 3D animations)
- To clarify the influence of and support from government and international organizations in relation to Afghan animation.
- To understand the challenges and opportunities for the future of Afghan animation industry.

The book is structured in three chronological sections;

1. Animation before 1996
2. New Phase of Animation (2001 to 2010)
3. Animation after 2010

From 1996 to 2001 (during the Taliban regime) there was a period of complete silence across cinema and animation in Afghanistan. In this period, cinemas were attacked or closed, many films were burnt, watching television and films was forbidden, and listening to music and creating moving images was banned.

The country's film industry entered a new phase in 2001 after the withdrawal of the Taliban, and the industry started further progressing 10 years later. However, cinema in Afghanistan has struggled to grow because of political instability and economic devastation.

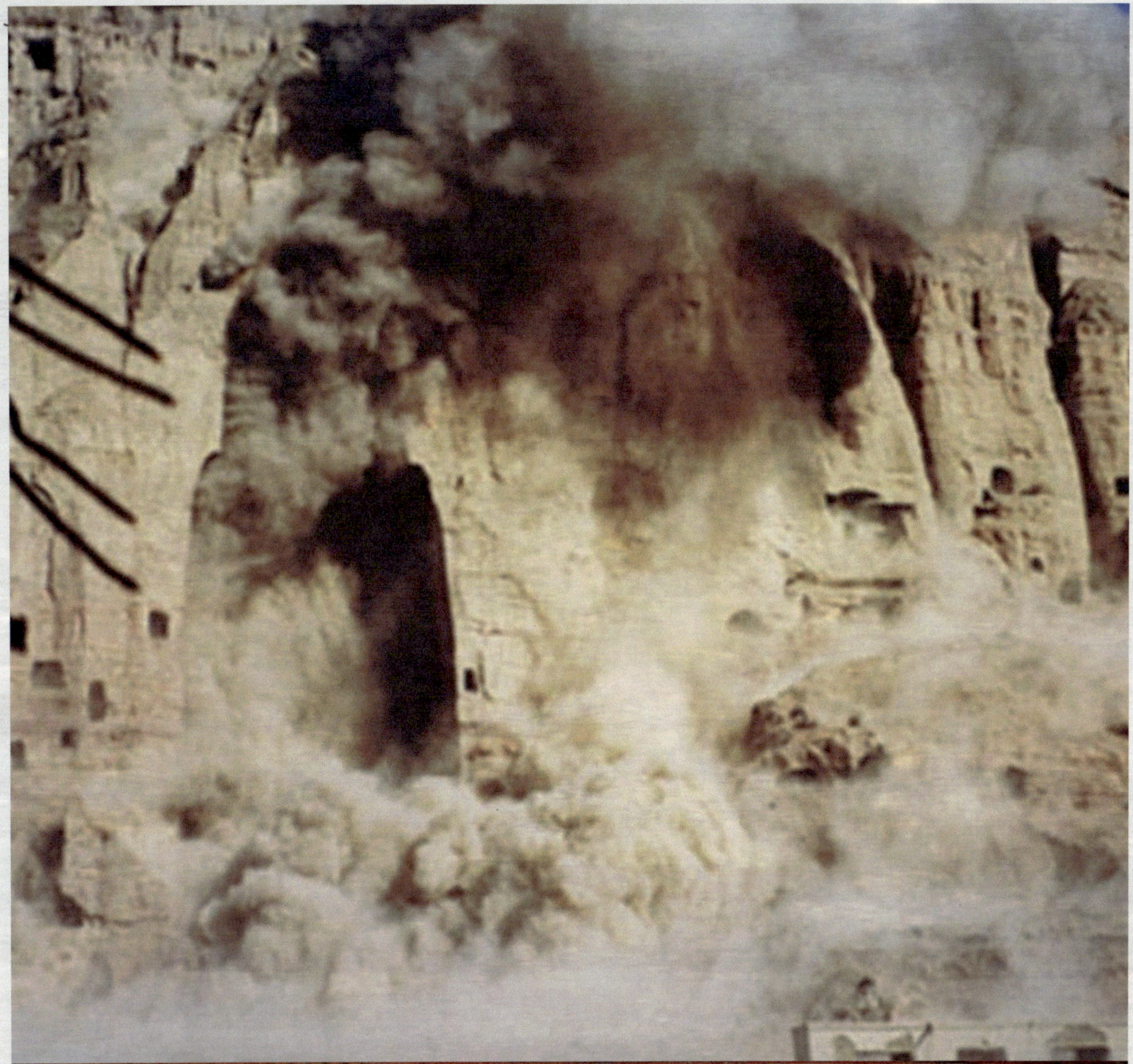

The destruction of Buddha statues in Bamiyan, Afghanistan 2001. Image source: Google

ABOUT AFGHANISTAN

About Afghanistan

Afghanistan is a landlocked country located in south-central Asia, which connects the Middle East and Europe to southern and eastern Asia.
The country's geographical location has made it a crossroads of cultures and empires throughout history.

Historically, Afghanistan's borders have shifted over time, with the ancient Silk Road passing through the region. The modern borders were established in the late 19th century. Currently it is bordered by Pakistan to the east and south, Iran to the west, Turkmenistan, Uzbekistan, and Tajikistan to the north, and China to the far northeast. Kabul is the capital of Afghanistan. Kandahar, Herat, Balkh, Bamiyan, Samangan, Mazar-e Sharif are its vibrant cities each with unique architectural and cultural landmarks. Both Farsi/Dari (Persian) and Pashto are the official languages of the country.

Since the beginning of the 20th century, many fine works of art of the pre-Islamic and Islamic periods were uncovered by archaeological researchers. The first ever oil paintings in the world were found in the caves near the two giant statues of Buddha of Bamiyan Valley in Afghanistan, which is 240 kilometers northwest of Kabul the capital. The paintings are estimated to be from the 7th century AD - hundreds of years before Europeans used oil paint. Samples indicate the oil paints from walnut or poppy were used in the paintings discovered by scientists in 12 of the 50 caves. "This is the earliest clear example of oil paintings in the world, although drying oils were already used by ancient Romans and Egyptians, but only as medicines and cosmetics", said Yoko Taniguchi.

The European Synchrotron Radiation Facility (ESRF) in France says "it was not until the 13th century that oil was added to paints in Europe and oil paint was not widely used in Europe till the early 15th century.

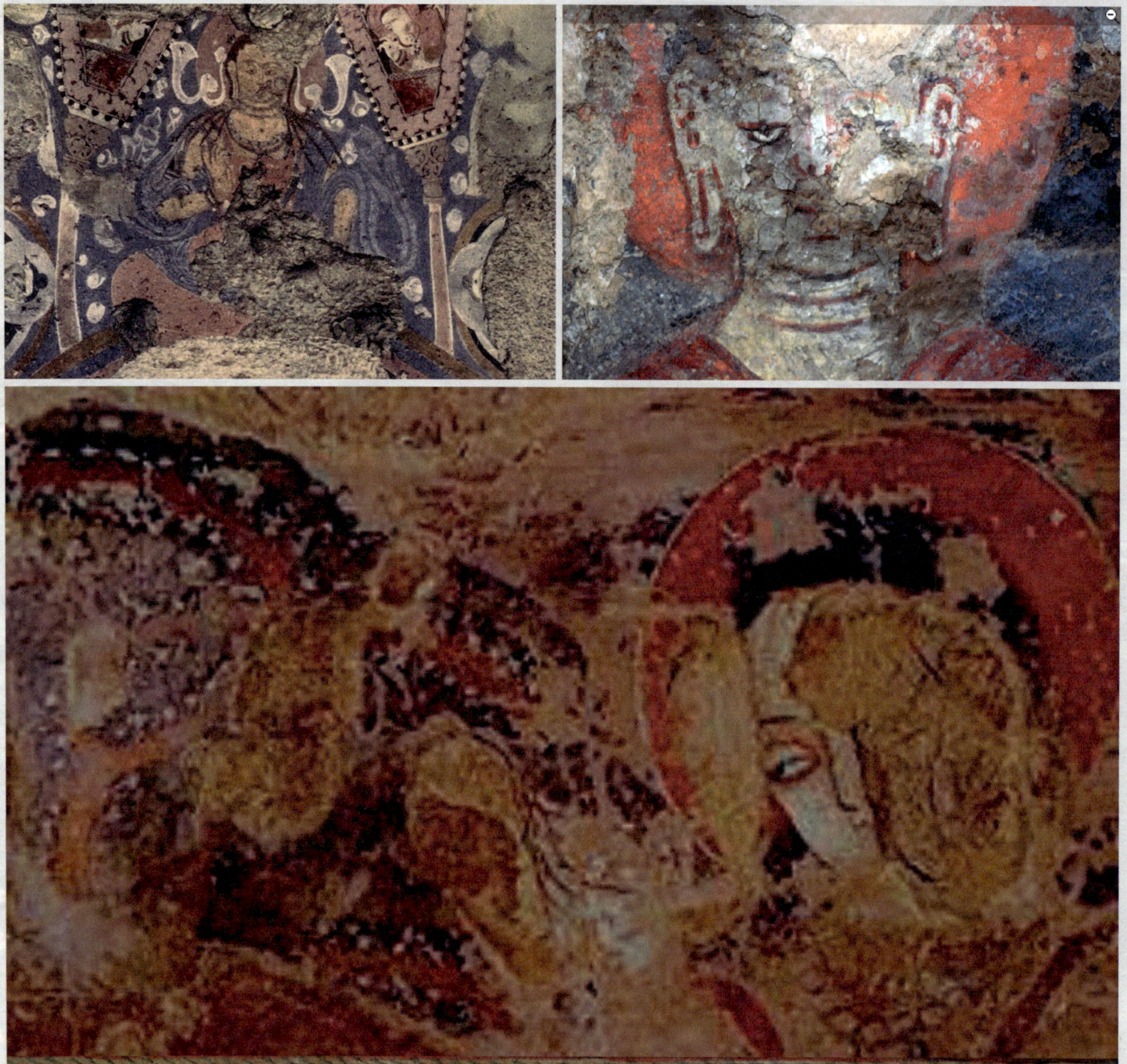

The first ever oil paintings in the world found in the caves of Bamiyan in Afghanistan. Image source: Google

Bamiyan was once a thriving Buddhist center where monks lived in a series of caves carved into the cliffs by the two statues. The cave paintings were probably the work of artists traveling along the Silk Road, the ancient trade route between China, across Central Asia to the West and show scenes of Buddhas in vermilion robes and mythical creatures".

Unfortunately, the caves were blown up and destroyed by the Taliban in 2001.

The civil war in Afghanistan (1978-2001) had a devastating impact on the country's art and culture, including:

- Destruction of cultural heritage: Many historical sites, museums, and libraries were looted, damaged, or destroyed, resulting in the loss of priceless artifacts and cultural treasures.
- Loss of artists and intellectuals: Many artists, writers, and intellectuals were forced into exile or killed, leading to a brain drain and a significant loss of cultural talent.
- Restrictions on artistic expression: The Taliban regime (1996-2001) imposed strict rules on art, music, and literature, banning many forms of creative expression.
- Damage to infrastructure: The war damaged or destroyed many cultural institutions, including theaters, galleries, and music venues, leaving a legacy of physical and cultural destruction.
- Displacement and refugees: The conflict led to massive displacement and refugee crises, causing many artists and cultural practitioners to flee the country, disrupting cultural traditions and community cohesion.
- Economic devastation: The war devastated Afghanistan's economy, making it difficult for artists and cultural institutions to access resources, funding, and support.
- Cultural erasure: The Taliban's attempts to erase Afghanistan's cultural heritage, including the destruction of the Bamiyan Buddhas, aimed to eliminate the country's pre-Islamic past and cultural diversity.

- Resilience and rebirth: Despite these challenges, Afghan artists and cultural practitioners have shown remarkable resilience, adapting and innovating in response to the conflict, and contributing to the country's' cultural rebirth.

Despite these lasting impacts, the country's rich cultural heritage and the determination of its artists continue to inspire hope and creativity.

In December 2001, after the Taliban regime collapsed, the media of the country came back to life. However, the term "animation" was still unknown among many artists and the public.

In 2010, however, with help from Goethe Institut, animation was formally introduced. The Institut conducted three rounds of animation training workshops. These workshops contributed greatly to the introduction and development of animation in Afghanistan. Numerous animation short films were produced during these workshops. At the end of third round of workshops, Goethe Institut held the first ever animation festival, the "Afghanimation-Animation Film Festival". Soon after, individual animators started producing and displaying their animations in different film festivals in the country.

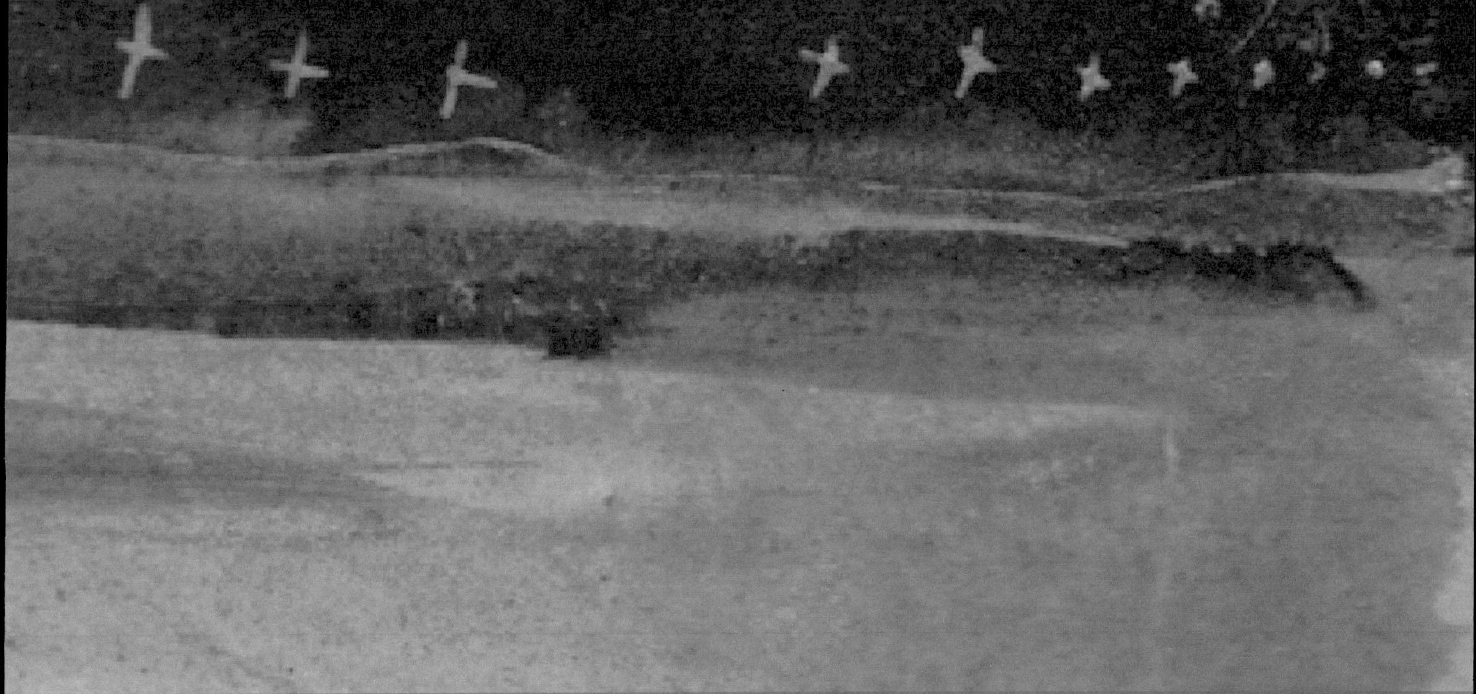

ANIMATION BEFORE 1996

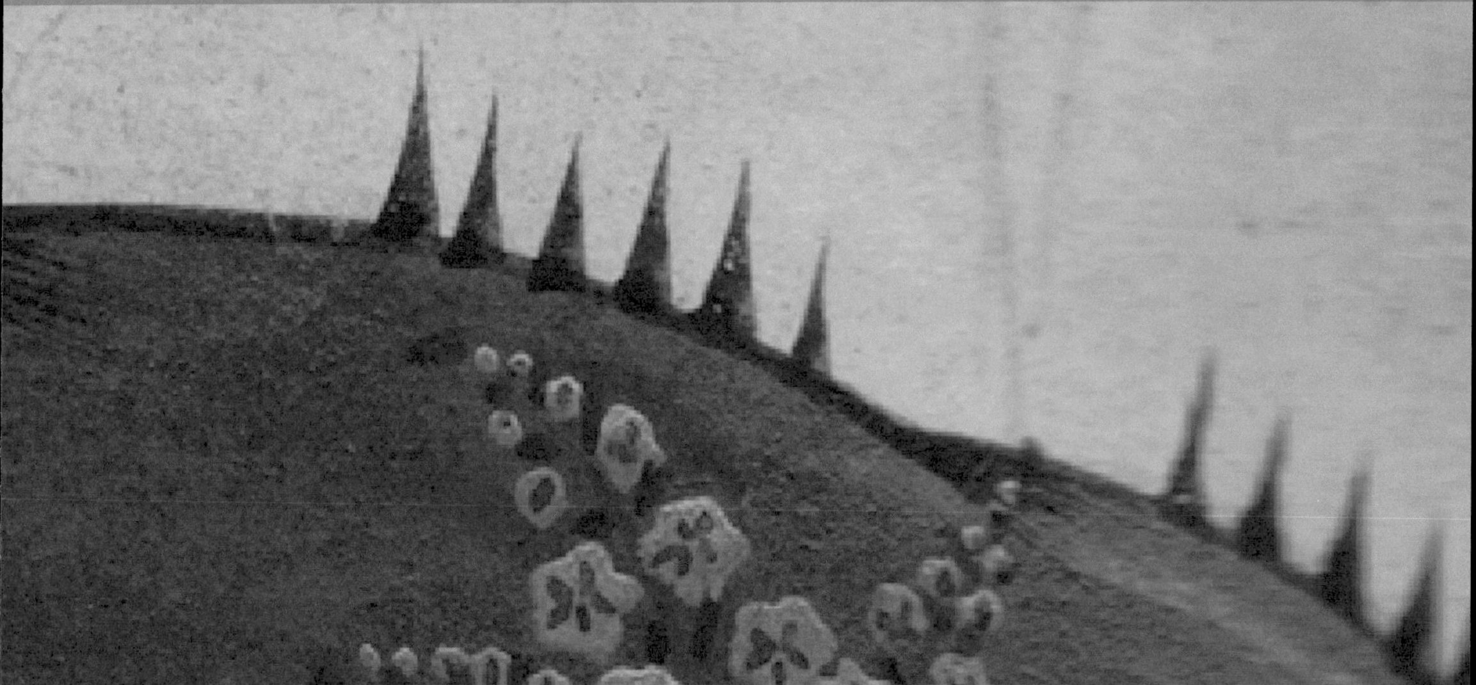

Animation Before 1996

Afghanistan has a rich history of art, with various forms of arts having emerged throughout history. From the 1st to 7th centuries, Gandharan Art flourished in Afghanistan. This form of art often features intricate carvings and sculptures and is characterized by depictions of the Buddha and other mythological figures. Moreover, Buddhist images painted in the caves of Bamiyan, Afghanistan are the world's first oil paintings and date to around 650 AD, These murals are the earliest examples of oil used in art history. Organic substances such as natural resin, plant gum, dry oil and animal protein as a binder have been used in these oil paintings. In the murals, thousands of Buddhas in vermilion robes sit cross-legged, sporting exquisitely knotted hair. Other motifs show crouching monkeys, men facing one another or palm leaves delicately intertwined with mythical creatures.

Islamic art is another type of art which emerged in the 7th century in Afghanistan. It is characterized by geometric patterns, calligraphy, and floral motifs. Miniature Islamic Art and Calligraphy is another style, which dates back thousands of years to the Sumerian time. It is characterized by intricate and detailed miniature paintings and calligraphy.

Realism Art was introduced to Afghanistan in the early 20th century by King Amanullah. It focuses on realistic depictions of people, places, and things. Other forms of art such as Impressionism, Modern Art, Fine Art, Performance Art and have also been used by Afghan artists.

In the 15th century, the Herat School was famous for miniature paintings in the Timurid period (1370-1507). In the miniatures of this school, numerous figures in groups or individually are shown on various planes, one above the other using the entire picture area. The presence of figures and elements of scenery one above the other produced the effect of one appearing to be behind the other. The figures of the earlier Herat school are stylized tall and thin with oblong heads and pointed beards but are painted in a variety of positions.

Above all, they are animated. Later, the Herat School was dominated by the figure of "Behzad".(Dupree, Afghanistan, 2020).

Kamal ud-Din Behzad (1455-1536) is a master miniature painter who was born in Herat, Afghanistan. He was raised and completed his studies under the supervision of Mirak Naqqash who was a painter in the Timurid period. Behzad's style has had a significant influence on the development of Persian art. His human figures were less stiff in their stances and more dynamic in their movements, creating a greater sense of energy and emotions to the paintings. In 1486, he became head of the Herat academy, a post he held until 1506. Under his direction, the academy became a greater center of art than ever before (Britannica, 2016).

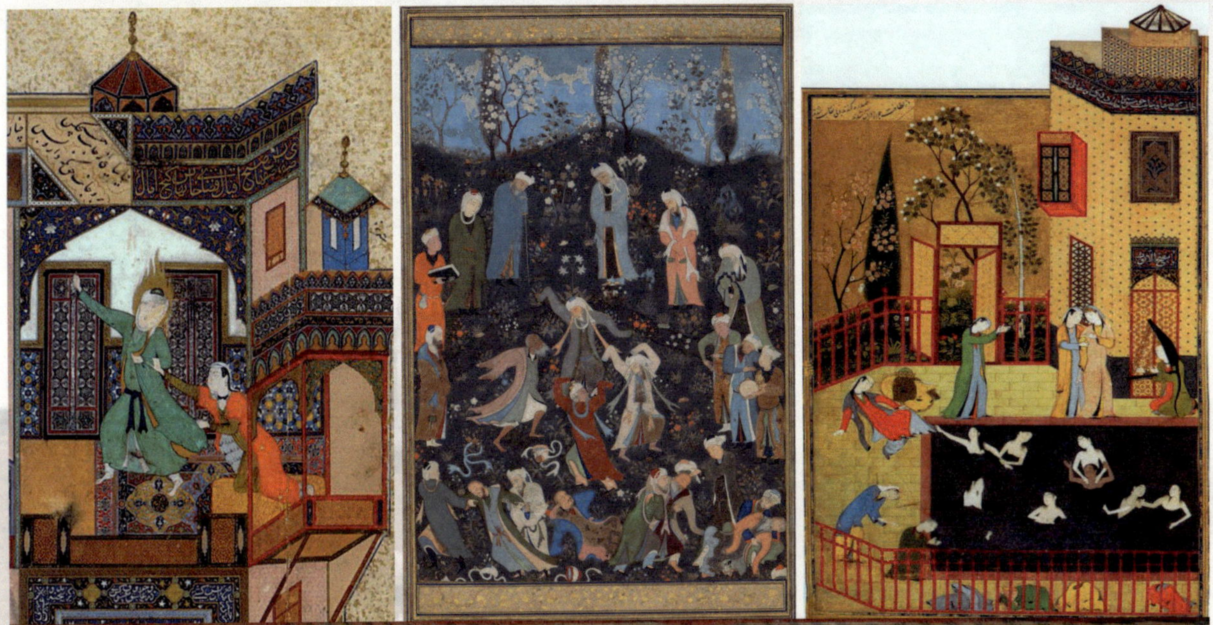

Artworks by Kamal ud-Din Behzad Afghan master miniature painter

During Shah Ismail Safavi's era, Behzad went to Tabriz, Iran. After Shah Ismail's death in 1524, he returned to Herat, Afghanistan. He died in Darul Saltanah, Herat in 1535 and was buried in Mukhtar mountains (Kôh-i Mukhtâr) Herat, Afghanistan. Unfortunately, his tombstone was partially destroyed and stolen, with the remaining piece being relocated to Herat Citadel (Qala Ikhtiyaruddin) for preservation.

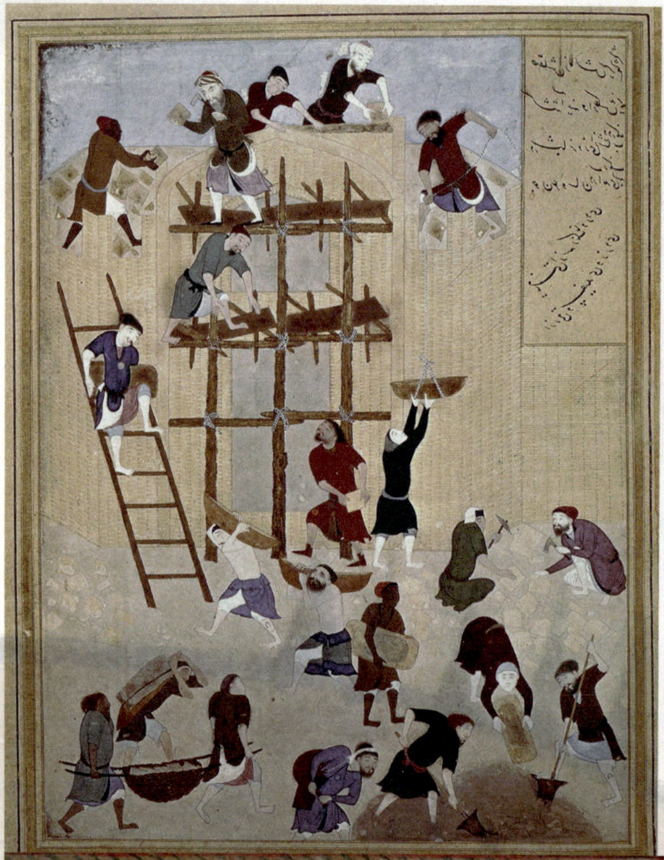

The Building of Khawarnaq Castle by Behzad, 1494

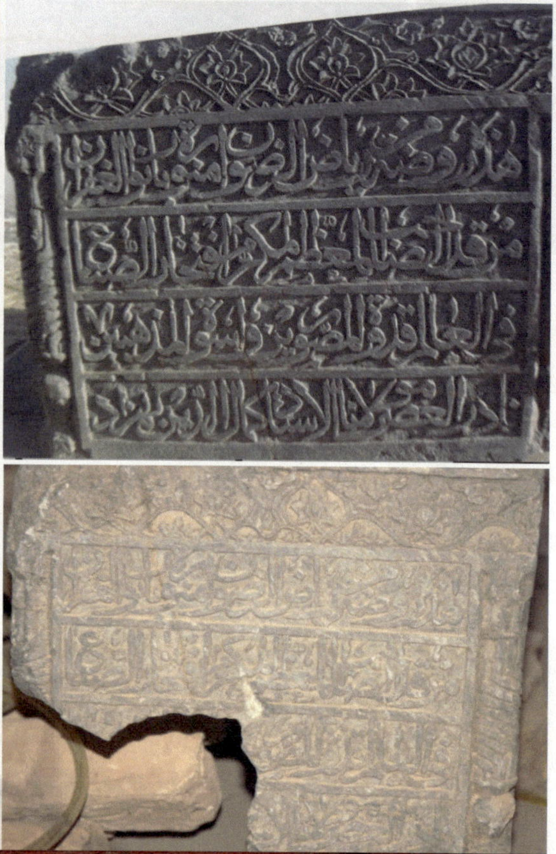

Carved stone from Behzad's tomb, Herat

Later in 1926, the country's second art academy, the Fine Arts "Sanaye Nafisa School" was founded by Ghulam Mohammad Maimanagi in Kabul. He was one of the best caricaturists of his time and his works were published in official gazettes of the time. He is considered one of the painters of the second golden age of realism in Afghanistan along with Abdul Ghafoor Brishna, Karim Shah Khan, Gholam Mahyeodin Shabnam Ghaznawi, Sayed Muqaddas Negah, Esrayel Roya, Abdul Rahim Nawin, Hafiz Pakzad and Akbar Khorasani who were creating art pieces and caricatures. This shows that Afghan artists were familiar with caricatures and motion. (Bayan, 2018)

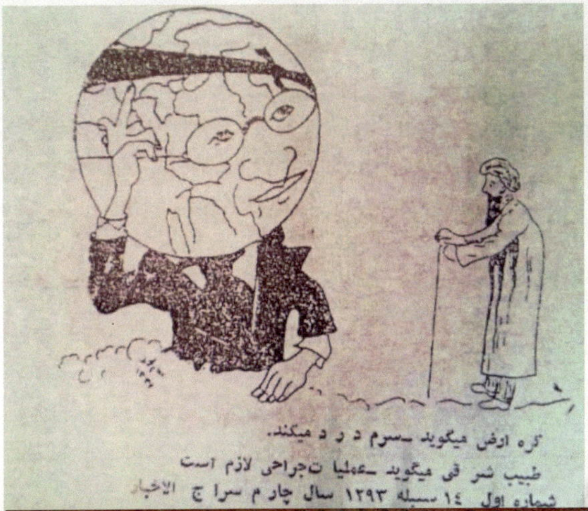

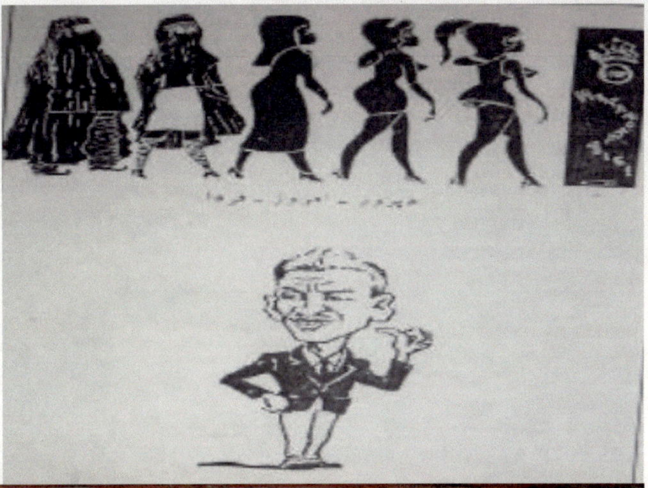

A cartoon by Ghulam Mohammad Maimanagi, 1914

A caricature by Abdul Rahim Nawin

Cinema emerged in Afghanistan at the beginning of the 20th century (1923). Amir Habibullah Khan, (who reigned from 1901–1919) introduced film to Afghanistan. The first projector "magic box" or "mageek lantan" (magic lantern) showed the first silent film to the public in Paghman palace in Kabul.

In 1946, the first Afghan movie, "Love and Friendship", was produced. Later, the first feature film 'Like Eagles' was made in Kabul by Afghan Film.

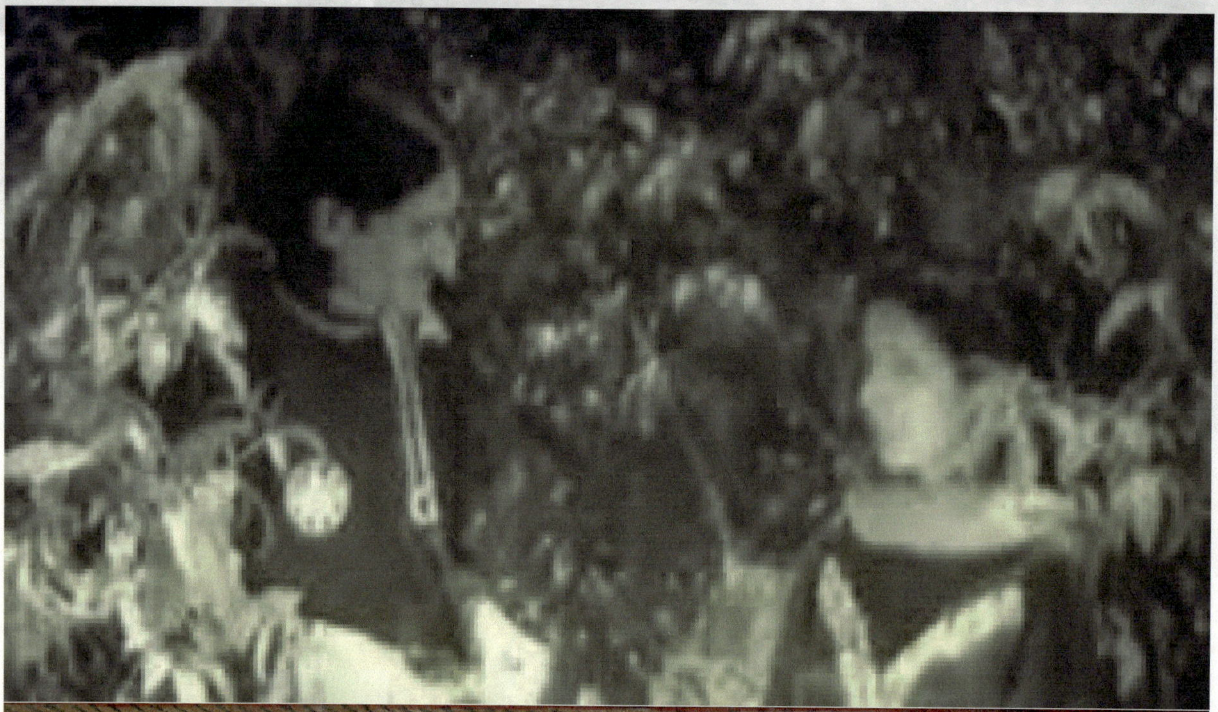

Love and Friendship (Ishq wa Dosti), Directed by Reshid Latif, 1946. Image source: mubi.com

Afghan Film produced a three-part movie with the collective title 'Ages', all of which were shot in black and white. The first colour films produced by Afghan Film in the late 1960s were 'Run Away' (Faraar), 'Love Epic' (Hamaasa e Ishg), 'Saboor the Soldier' (Saboor Sarbaaz), Ash (Khakestar), 'Last Wishes' (Akharin Arezo) and 'Migrating Birds' (Paranda Mohajer). (A Brief History of Afghans and Films, 2005).

Later, the movies (starting with 'Radio Kabul') were shared with the public through Radio Television of Afghanistan (RTA), founded by Amir Amanullah Khan in 1926. Construction works for a TV studio and transmission buildings were finished by August 1978 with the help of grant aid from Japan. This was the first TV station in Afghanistan. The station was often used by musicians to record music videos.

During the 1980s, along with movies, songs and other programs, many Soviet programs were airing such as the children's show "Nu Pogodi" (Russian animated series). Later RTA made a puppet animation "Gorg wa Khargosh" which was the only Afghan cartoon at that time and still its available in Archive of this Radio Television. Finally, the first animated film (Tamrin) was produced in 1987 (Sediqi, 2019).

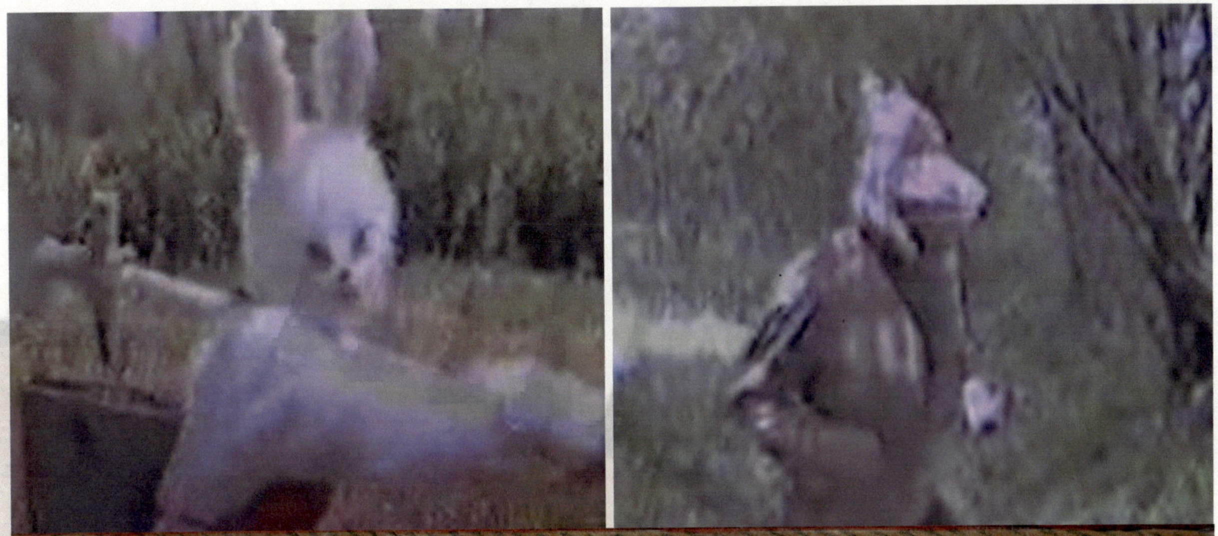

Screen shots of Gorg wa Khargosh Afghan Pupets, 1987. https://www.youtube.com/watch?v=L2Z9V48dXrw

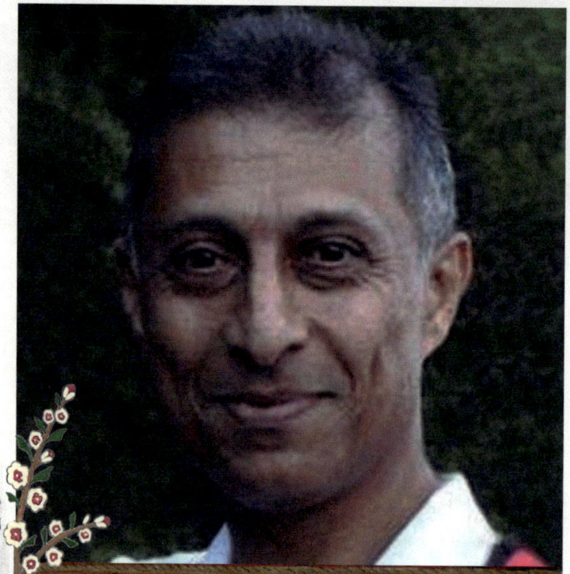

Hozhaber Shinwary The First Afghan Animator

The First Afghan animator

Mohammad Humayoon Hozhaber Shinwary, an Afghan caricaturist and animator, made the first two animations in Afghanistan. He started working on the first animated short film “Tamrin” (Exercise) in 1987, with a small team and help from National Radio Television and Afghan Film Organization. Later, he produced the second animated film, “Parwaz ba soye Khorshid” 1989 (A Flight Towards the Sun). These two films were animated using traditional techniques in which thousands of drawings were used. These two are the first animated short films funded and produced entirely by this team in Afghanistan. Unfortunately, these films are not available now neither in Archive of Afghan Film Organization nor in archive of National Radio Television of Afghanistan RTA.

I found an interview from 1990 on YouTube, broadcast by the National Radio and Television of Afghanistan RTA. In this program, Mr. Shinwary's animation team (Parween Pazhwak, Hassan Samedi and Jawad Mansour Popal are invited to talk about the animations they created. The interview reveals that the first Afghan cartoon film was "Tamarin" and the second film was called " Parwaz ba soye Khurshid ". The program also shows that these two films were shown in the Afghan Film Hall. Some parts of the short animations are also featured in this interview.

Parween Pazhwak says that the animation of “Parwaz ba soye Khurshid” is made in a symbolic way with elements such as flowers, sun, pigeons, and a girl. Each of the elements has its own specific meaning. She added that 1400 drawings are used in this film and its duration is five minutes. She also says that in the first film, “Tamrin” 10,000 drawings were drawn and then 5,000 were used in the film.

Hassan Samad, another member of the team, says that when he watched the film at the Afghan Film Hall for the first time, his interest in creating moving images grew. The team members recognise Mr. Shinwary as the artist, director and cinematographer of these animated films.

Jawad Mansour Popal introduces the characters of their next film and adds that the scenario of the next film is finished, the characters are ready, and they want to start filming. Although he does not say the name of the film, he says it will be about the life of a musician. These interviews are evidence that both of the animated films were broadcasted by the National Radio and Television of Afghanistan at that time.

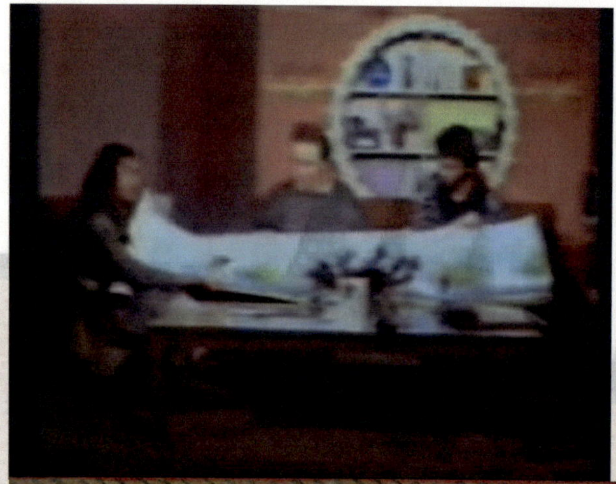

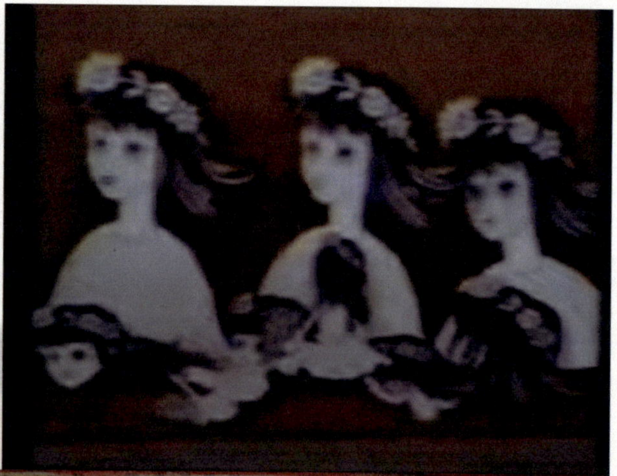

Screen shots of Animation team with RTA, 1990. Link: https://www.youtube.com/watch?v=MJLkq72adCs

I had an online interview with Jawad Mansour Popal in June 2024 and he added the success of the project was a result of the collective efforts of 10 individuals including himself, listing the following team members:

1. Humayoon Hozhaber Shinwary
2. Said Hassan Samad
3. Parween Pazhwak
4. Dr. Yousuf Asifi
5. Nilab Pazhwak
6. Nilab Salaam
7. Luna Salaam
8. Asad Badi
9. Zulaikha

The political changes in Afghanistan over many years did not allow the film industry in Afghanistan to grow. When the Taliban were ruling the country from 1996 to 2001, cinemas and theatres were destroyed and remained closed even watching television, listening to music and creating moving images were forbidden. Lots of artists escaped Afghanistan. Afterwards, the film industry of Afghanistan entered a new phase in 2001.

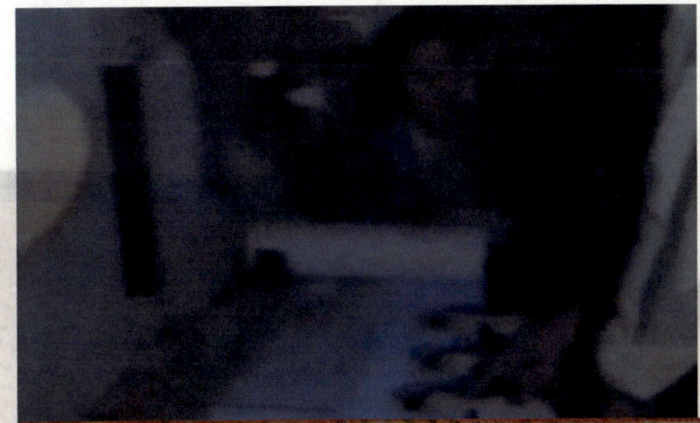

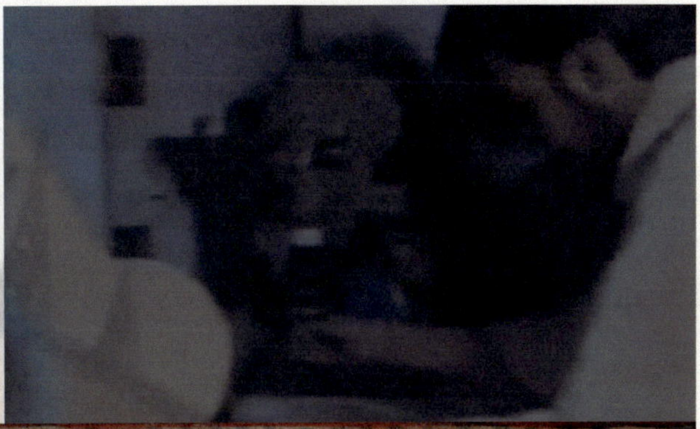

Screen shots of Jawed Popal interview with RTA. 1990 Link: https://www.youtube.com/watch?v=Jh5qYW3oQK8

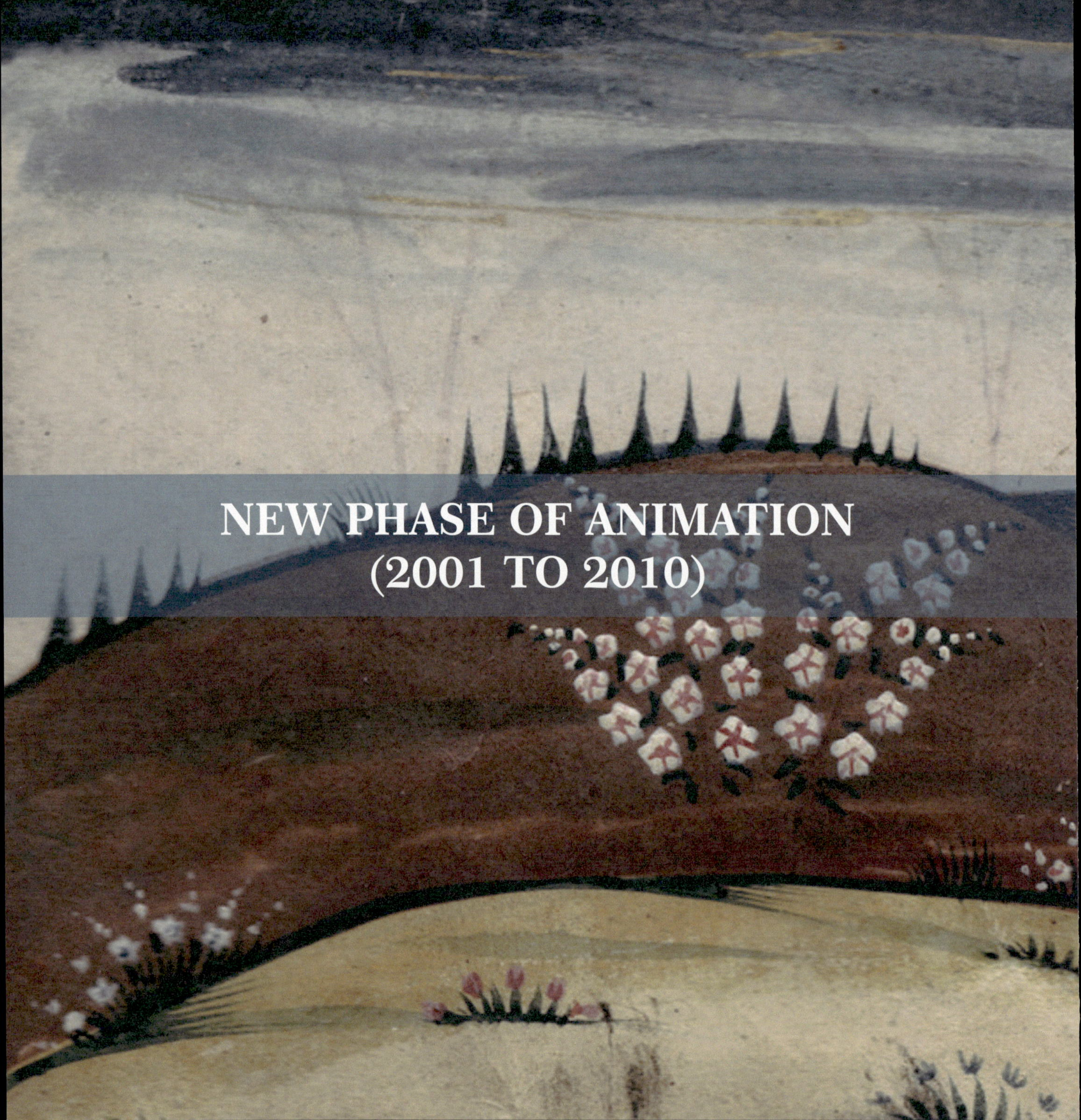

NEW PHASE OF ANIMATION
(2001 TO 2010)

The New Phase of Animation

Soon after the Taliban regime collapsed in 2001, anti-Taliban forces agreed on a period of transitional leadership and an administration that would lead to a new constitution and the establishment of a democratically elected government.

In December 2001, media in the country came back to life. Afghanistan's earliest television channel (RTA) was relaunched and Tolo and Shamshad were the first commercial TV stations in the country. Artists once again started creating modern art, and contemporary artists started experimenting with new styles and mediums, including digital photography, textiles, abstract painting, filmmaking, and mixed media. Performance art also emerged in Afghanistan particularly after the fall of the Taliban in 2001 - including musical puppetry and other forms of storytelling.

In 2002, FKH media started working as a production company in Afghanistan. Along with video production (production of documentary and fiction films, radio and television ads), FKH Media opened a specific section for animation, which was established three years later. The production company started working on a 3D animated series titled "Yasin and Kaka Rauf" in 2006, made for the USAID project 'Rule of Law'.

During this time, animator Mr. Mohsen Hussaini produced his first animated short film, "Shelter" which is about six minutes. This animation was made using traditional techniques in 2006.

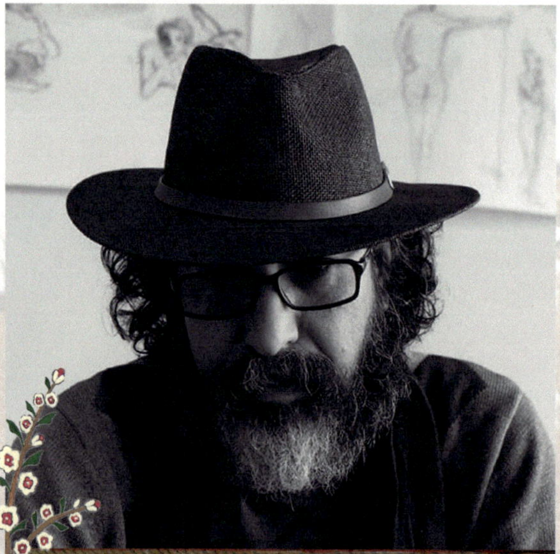

Mohsen Hossaini, The first animation instructor

First Animator of the New Phase, Mohsen Hossaini

Mr. Mohsen is one of the most well-known artists in Afghanistan. He is a writer, graphic designer, animator and filmmaker who has a very unique style throughout his works. Hossaini was born in Kabul in 1977. At the age of seven, he and his family moved to Iran where he completed his Bachelor degree in Animation from I.R.I.B University (Iran Broadcasting University) of Tehran and began producing animation. He returned to Afghanistan in 2005.

Since then, he has made six animated short films, entitled "Shelter", "Pefume", "Hitler", "Dance", "Red Fish and "Event" and "Earthquake". I interviewed Mohsen Hossaini for the second time in May, 2024 and he said he was working on a new animation "Border". His animated films have been screened in different national and international festivals across the world, including Hiroshima, Leipzig, Bosnia, Krakow, Cologne, and Stockholm. He is the first animation instructor in Afghanistan who has trained 30 students through animation workshops, supported by the Goethe Institut in Kabul. Through these efforts, he has opened a path for other animators in Afghanistan. Furthermore, he has worked as professional graphic designer for nearly 19 years in different media companies and has exhibited his paintings and caricatures in different exhibitions. He has also five documentaries in his records.

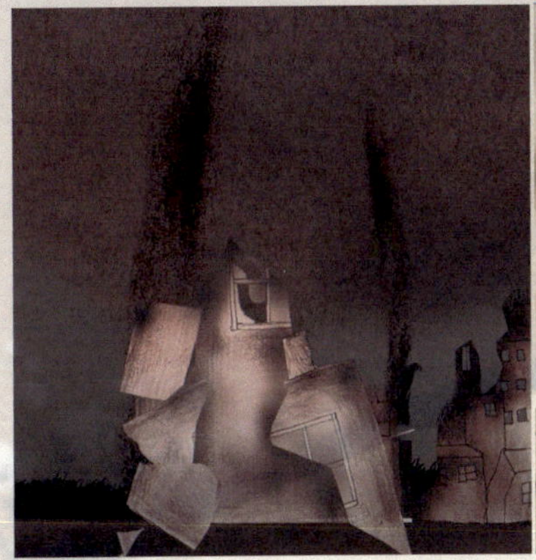
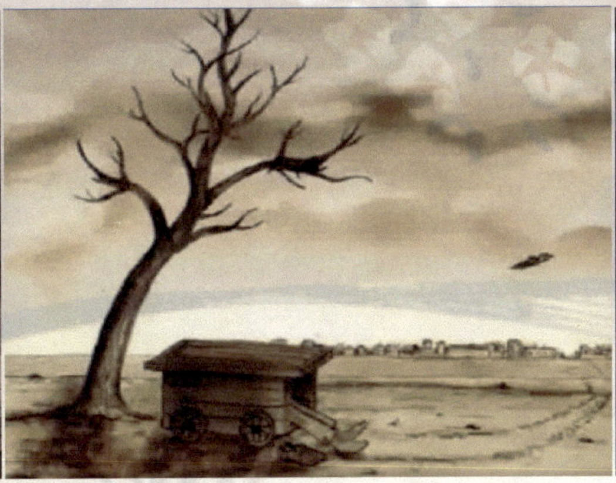
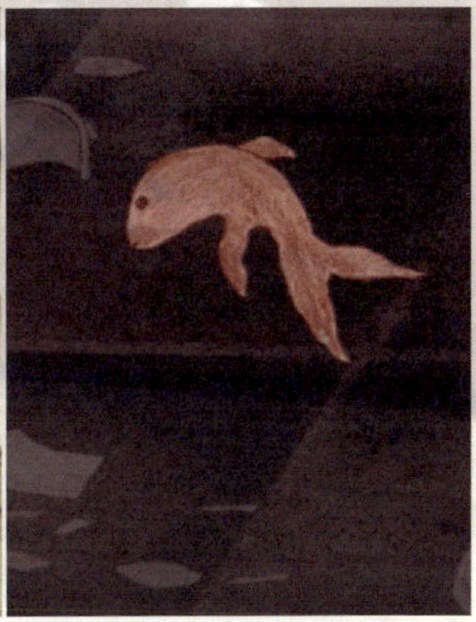

FKH Media, Production Company

FKH Media was established in the year 2002 in Kabul as a production company with high quality equipment and an experienced team of national staff. It has been one of the leading companies providing high quality media products and public outreach services in Afghanistan.
This company is involved in the three following fields:

- Producing media products
- Designing and launching advertisements and public awareness campaigns
- Technical services and providing rental equipment to filmmakers

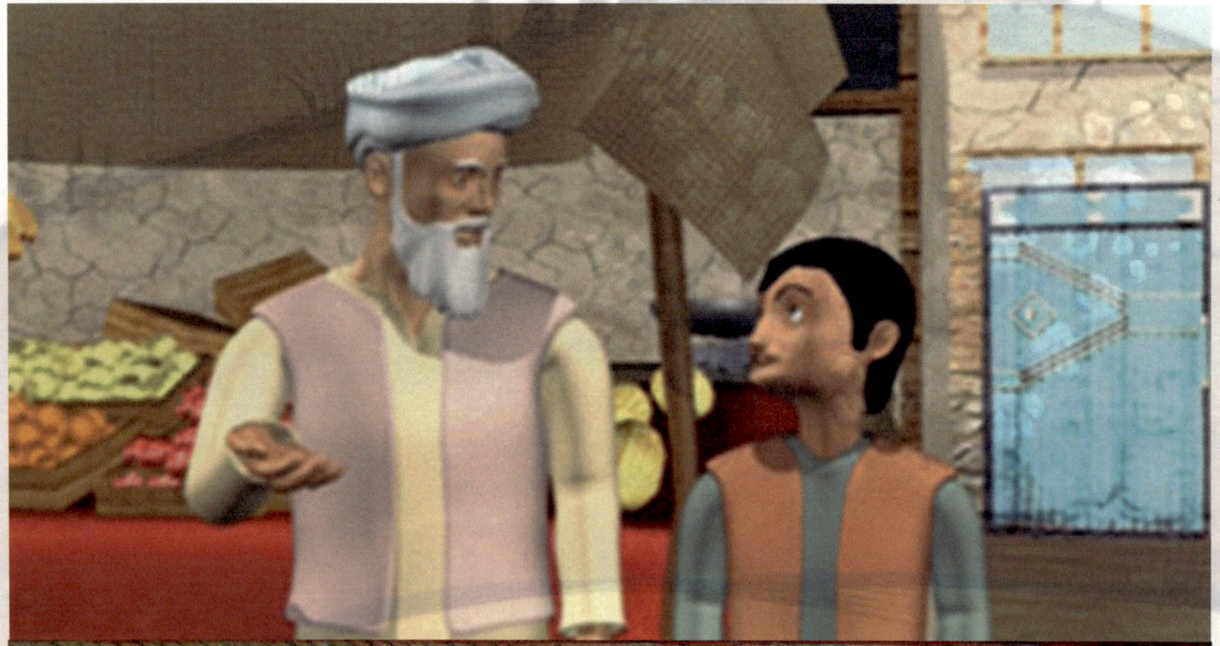

Yasin and Kaka Rauf 3D Animation 2006

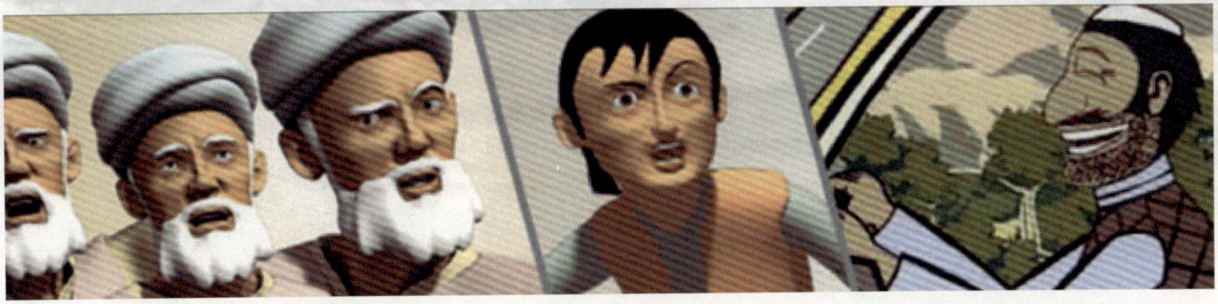

Apart from video production, the company also produces documentary, fiction films, and radio television ads. FKH media opened a specific department for animation in 2005. A year later, it started working on a 3D animated series titled "Yasin and Kaka Rauf". The company successfully produced the first 3D animation series in Afghanistan for the USAID project "Rule of law", based on collection of six comic books which was made by Sayara company. A group of four worked to produce these series of animations. Each episode is around four to six minutes. Each episode was completed in three to five months which were later aired on national televisions. Four of the six episodes are made in 3D, while two episodes are in 2D. In 2013 and 2014, the seventh and eighth episodes of this series were produced by the same production company. The software used in this series are Toon Boom, 3D Max and Maya. The series was formally inaugurated at the Ministry of Information and Culture, attended by representatives of USAID, the Directorate of Afghan Films and other organizations. This event was given big coverage by the national and international media.

Later, FKH media, developed the skills, techniques and equipment toward creating & producing more quality animations. Currently, they use these techniques in order to create visual effects in Movies & TV Ads. Moreover, FKH Media has dubbed 104 episodes and 1560 minutes of the series "Postman Pat" in two languages of Dari and Pashto (a British animated series for children). This was sponsored by the social programs of Roshan Telecommunication company.

Kabul University, Faculty of Fine Arts

Kabul University is the oldest and the largest university in Afghanistan. The faculty of Fine Arts is one of the many faculties in this university. This faculty was founded in 1966 and considered to be one of the most important centers of arts in the country.

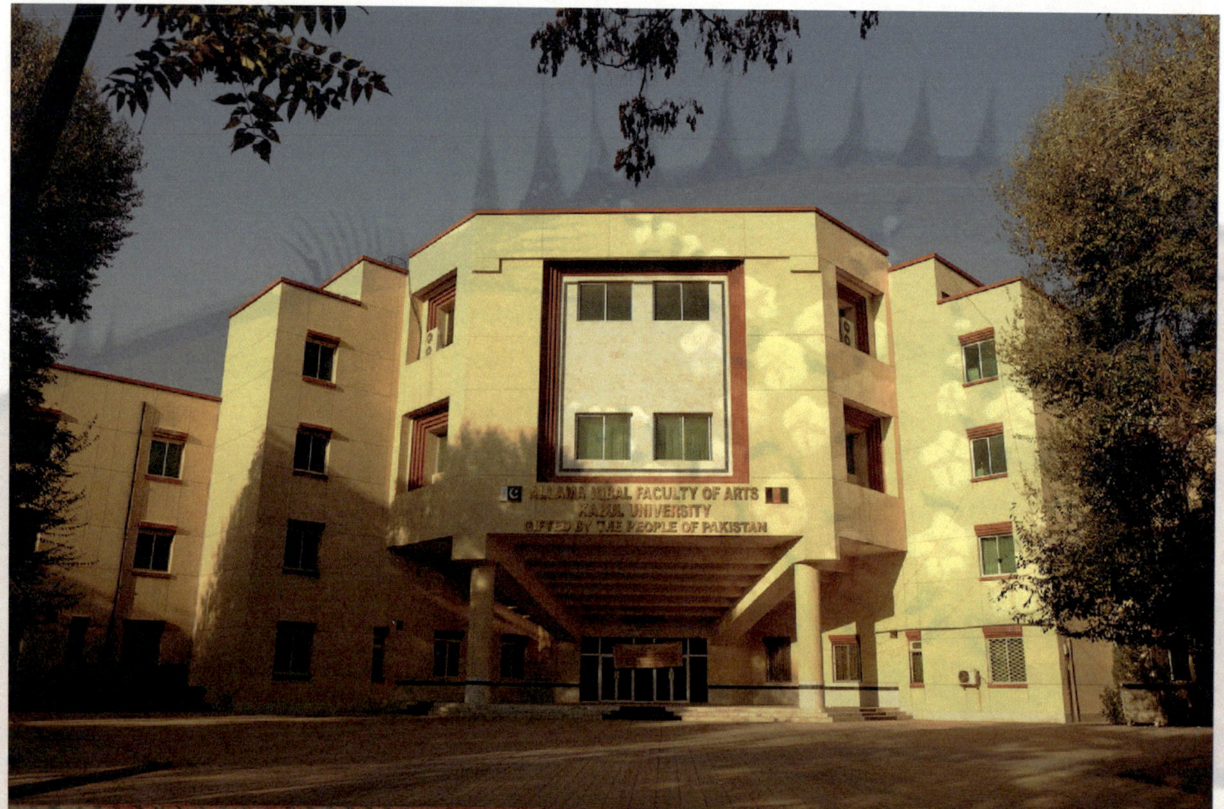

Kabul University, Fine Arts Faculty, 2019. Iamge by Aziz Azizyar

In 1966, Professor Amanullah Haidarzad, completed his Master's degree in the field of sculpture and oil painting from Italy. After his return to Kabul, he started working in the Kabul University and he established academic courses in sculpture and painting within the Faculty of Literature and Human Sciences. In 1968, in addition to the classes of painting and sculpture, classes in music, acting, photography and calligraphy were added within this faculty. This included eastern music, which was taught by Mohammad Hossein Sarahng, and western music, which was taught by German professors.

Mr. Latifi and Mr. Abdul Hamid Fayaz collaborated in the fields of allegory, photography and journalism. Finally, through the efforts of the professors, the Faculty of Fine Arts was officially accepted as a department of Human Literature Sciences. Professor Amanullah Haidaerzad was appointed as the Head of the Department and later it was recognized as an independent faculty. In 2021, the Faculty of Fine Arts had eight departments which were as follows

1. Department of Sculpture
2. Department of Painting
3. Department of Graphic Design
4. Department of Cinema
5. Department of Music
6. Department of Theatre
7. Department of Photography
8. Department of Dramatic Literature

The Departments of Sculpture and Music were removed by the new regime in 2021 and replaced with a Decorative Arts department. However, because many academic staff fled the country, most Departments were without professors. At the time of writing this, only four of the Departments (Graphic Design, Cinema, Theater and Painting) have students. Unfortunately, a ban has been imposed on the education of girls and the activities of female professors, who are not allowed to enter the university and study.

I interviewed Mr. Samiullah Nabipoor the head of Fine Arts Faculty, on 25 January 2020 in Kabul University. Mr. Nabipoor said that he was planning to initially make animation part of the Cinema or Graphic departments. But, unfortunately, this did not happen, and animation never became part of the academic structure.

I also visited the Kabul University library and conducted an interview with the General Director of Central Library of Kabul University, Mr. Habib Ullah Lemar "Jahanyaar" on 25 January 2020. Mr. Jahanyaar added that, "there are many books in the library which are related to different fields but there is no book available in Animation." To date, no book has been written on Animation by Afghan authors. Only one thesis is available in this library which is about animation. This thesis was done by Rafi Bayan who is an Afghan artist".

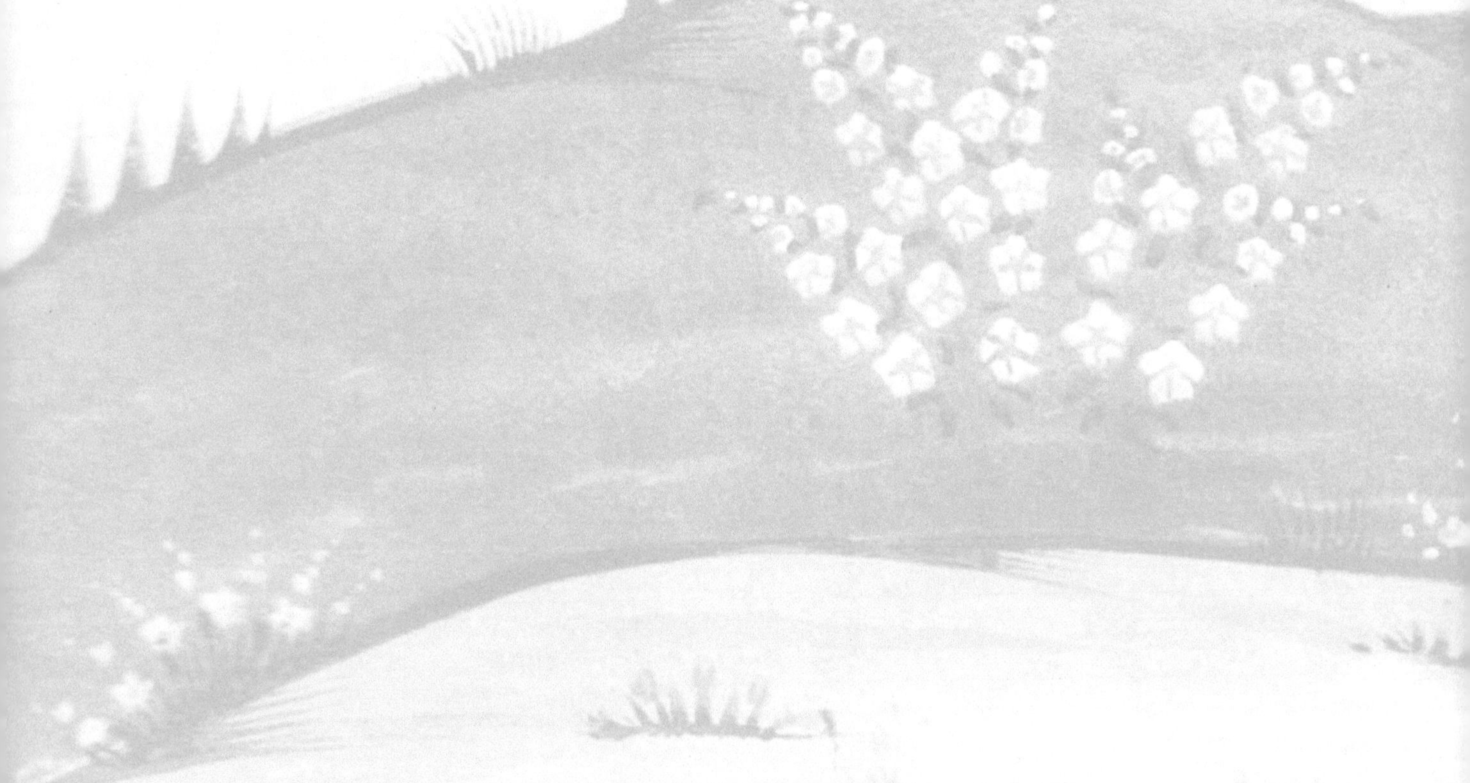

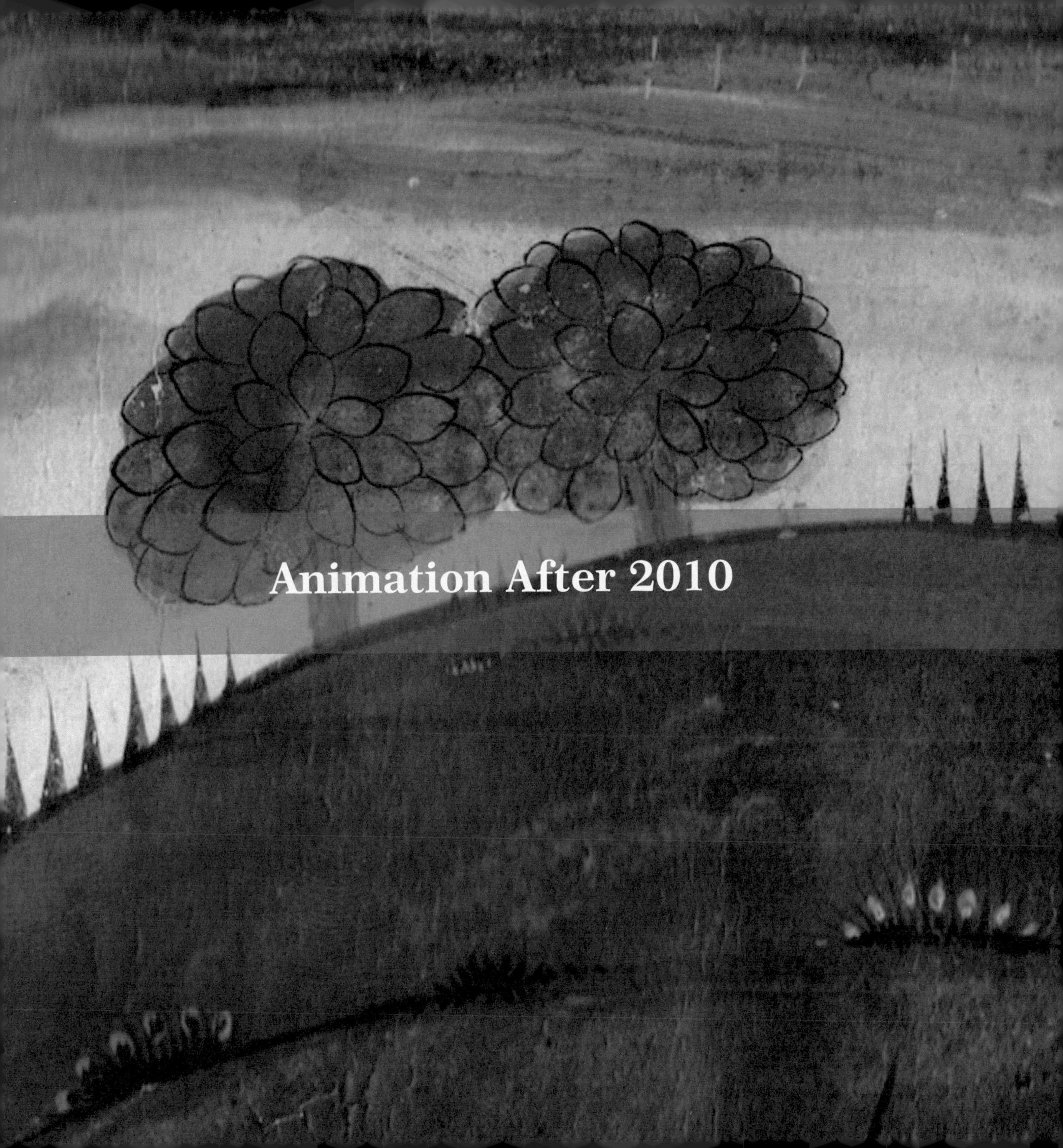

Animation After 2010

Animation After 2010

The term "Afghan Animation" has been largely unknown in Afghanistan. even among artists. However, in 2010, there was a revolution in the history and development of animation in Afghanistan. Globally Active Cultural Institute of the Federal Republic of Germany (Goethe-Institut), started a project to support the growth of animation in Afghanistan. Mr. Mohsen Hossaini, a professional Afghan animator, was in charge of the training workshops conducted by this institute. From 2010 to 2013, three rounds of educational workshops were conducted. These workshops were the first formal academic animation trainings in Afghanistan. A number of animated short films were produced by the students of this workshop. At the end of third round of the workshops in 2013, the institute held a festival named "AfghAnimation-Animation Film Festival" - the first ever animation festival in Afghanistan.

Moreover, in 2011, the first animation movie Buz-e-Chini was made and released, marking another milestone in the history and development of animation in Afghanistan. By the end of this year, "Baghch-e-Simsim" (a children's televisions series based on the American Sesame Street) was also launched in Afghanistan.

Soon after, individual animators started producing their animations. Different festival holders in the country tried to encourage animators by screening the produced animations in the film festivals. Gradually, animation became more renowned and formed part of media and advertising in Afghanistan. However, animation did not progress much in Afghanistan due to security and political instability.

The Goethe-Institut in Afghanistan

The Goethe-Institut in Kabul, Afghanistan, has been a beacon of cultural exchange and understanding since its reopening in 2002. This esteemed institution has been dedicated to facilitating cultural contacts between specialists, and fostering a rich tapestry of artistic expression. Through its diverse activities, including film, music, theatre, painting, German language classes, and library support, the Goethe-Institut has created a vibrant platform for Afghan and German artists to come together, share ideas, and learn from each other. This enduring cultural bond has been strengthened by the institute's commitment to promoting cross-cultural understanding and exchange, making it a shining example of the power of art and culture to bring nations together.

In 2010 the Goethe-Institut initiated an animation project in Kabul, Afghanistan. The Institut provided workshops and training programs for Afghan artists, empowering them to tell stories and showcase their talents by creating animated content.

The main objective of Goethe-Institut's educational workshops led by Mr. Mohsen Hussaini was to teach budding animators how to direct an animated movie or short film. Mohsen Hossaini helped the students learn all preproduction, production and postproduction stages and students were able to make animated short films at the end of the course. Each round of the workshops took around four to five months to complete. In each round, ten students were selected and trained.

Mr. Mohsen Hossaini, animator and Mr. Ibrahim Hotak, head of Goethe Institut in Kabul, played vital roles in delivering these programs, however due to security problems in Kabul, Goethe Institut was closed in 2018 and the animation workshops and film festivals have no longer taken place.

Ibrahim Hotak, Head of Goethe-Institut in Kabul

Mr. Ibrahim Hotak was born in Kabul in 1963. He studied arts, philosophy and culture management in Germany. He has a B.A in German language and can translate across German, Dari and Pashtu. Mr. Hotak was head of Goethe Institut in Kabul and worked to strengthen artistic cultural development in Afghanistan. He is the person who created a platform for the first generation of animators to learn animation through Goethe Institut workshops and helped them to project their animations through the Afghan Animation Film Festival.

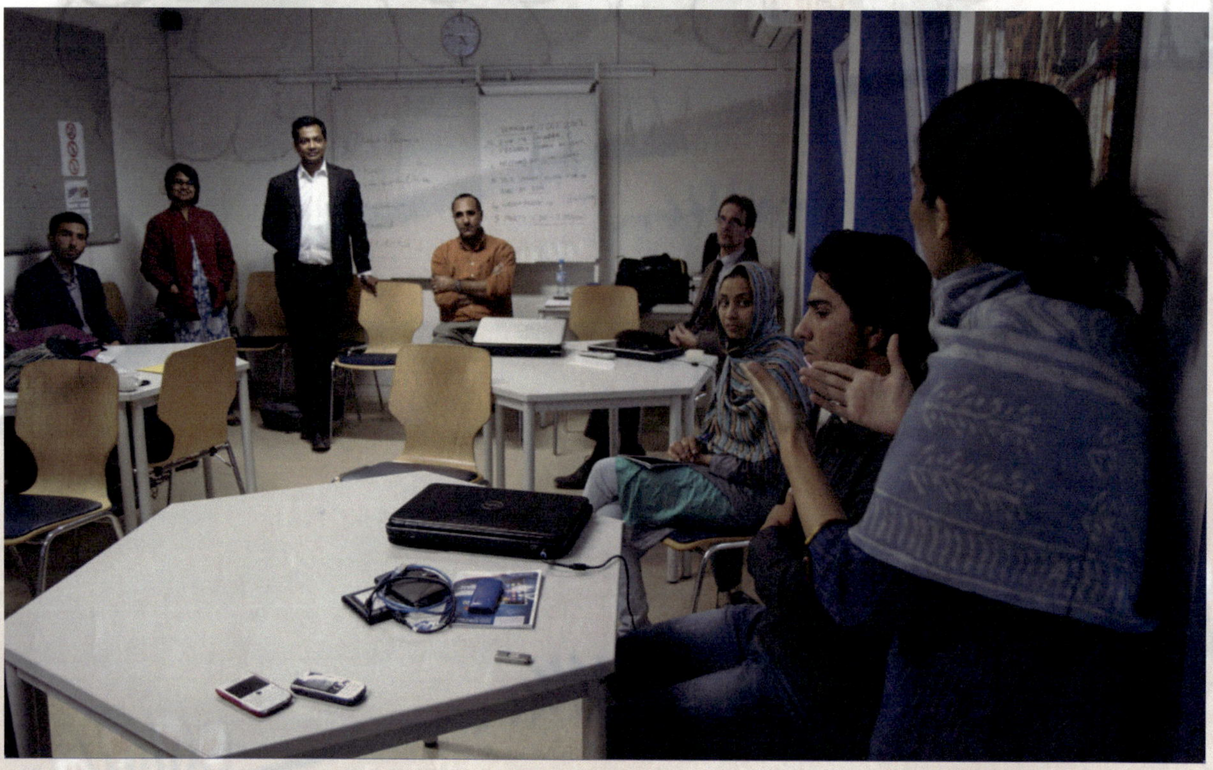

AfghAnimation-Animation Film Festival

This Festival was organized and held by Afghan animators such as Mohsen Hossaini and Nasir Hashimi with financial support from Goethe Institut and with the help of Mr. Ibrahim Hotak, Head of the Institute. The first round of these festivals was held in 2013 in Kabul. In this round, 26 animated films by 18 animators were screened in Faculty of Fine Arts' Hall, in Kabul University.

In the second round of the festivals which was held in 2014,15 animated films were screened from 12 animators in the same location. The third festival was held in Kabul University auditorium in 2015 but due to the participation of only a small number of animations, it was made a joint festival for animation, short fiction and documentary which was titled "The AfghAnimation Short Fiction and Documentary Film Festival". It included four animated short films, one documentary and 22 short fiction films.

In 2015, with help from Goethe Institut, animation became a department in Afghan Film Organization. The head of Animation department of Afghan Film, Mr. Nasir Hashimi continued conducting animation festivals and the fourth round of the animation festival took place in 2017 in which ten short fiction films along with 15 short animated films were screened in Kabul University. Kabul University representatives have encouraged and welcomed Afghan Animation Festival holders and young artists by providing them a venue to display their animations and hold festivals. When Goethe Institut was closed due to security challenges, the animation film festivals were no longer held.

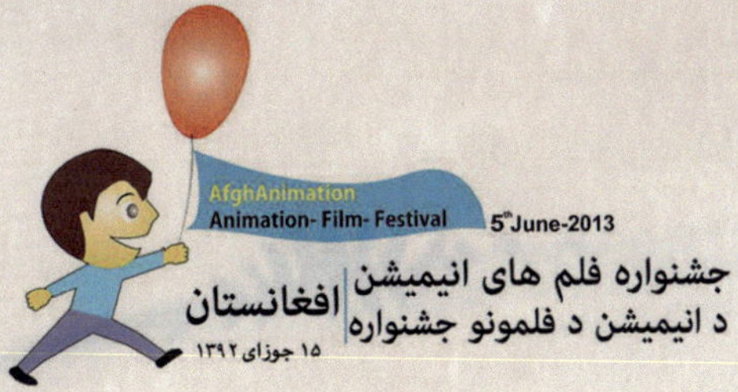

Afghan Film Organization

The Afghan Film Organization (AFO) is one of the directorates of the Ministry of Information and Culture of the Afghan government. In January 2020, I visited AFO's office in Kabul and conducted an interview with the Director Ms. Sahraa Karimi, to learn about the organization and their activities in animation.

In this and other interviews with Mr. Nasir Hashimi (former Head of Art, Graphic and Animation department of this organization; and Mr. Hizbullah Sultani, present head of this department), I learned that the organization had an animation department which opened in 2015 with financial support from Goethe Institut.

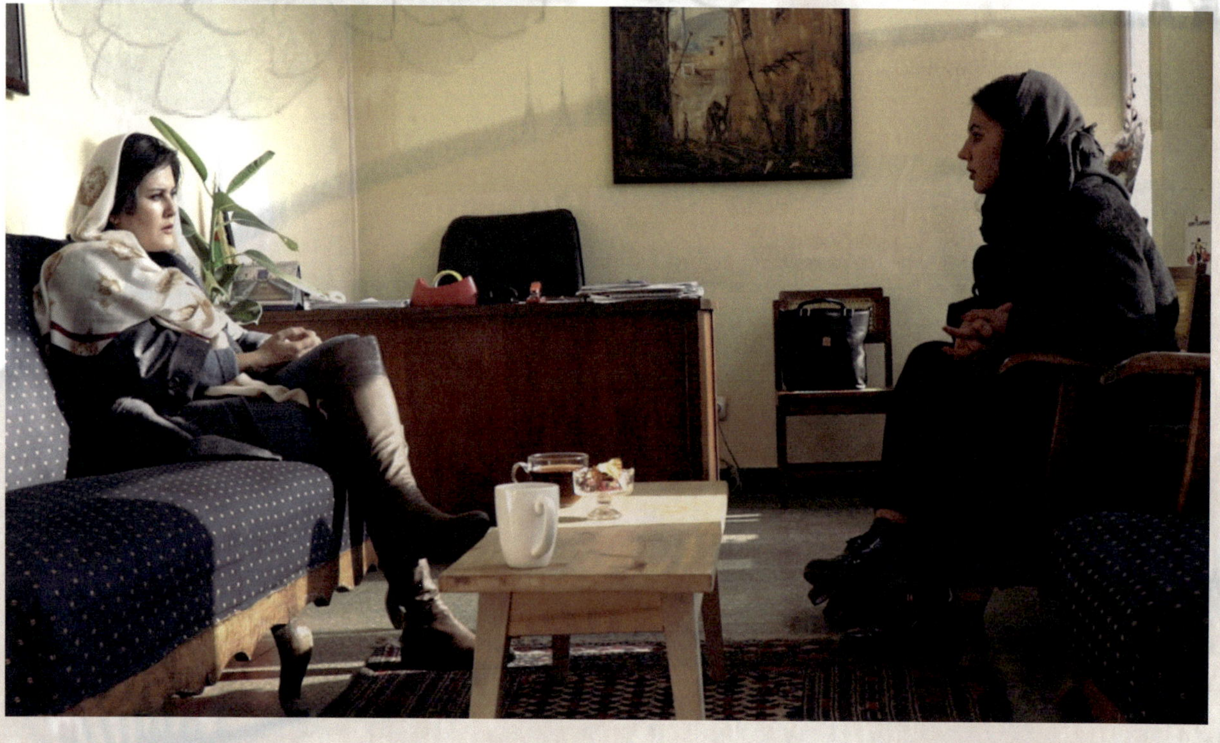

When this department began, the head of the department Mr. Nasir Hashimi, with the help of Hizbullah Sultani, collected information on almost all animations made in Afghanistan; mainly, from 2010 to 2016. They found 61 animated short films and created an archive which includes various types of animations and topics from male and female animators.

The head of this department at that time, Mr. Hizbullah Sultani, gave a copy of this collection to me for research purposes. This helped me prepare a list of Afghan animators and the animations made in the country, and to write about the animation techniques used. In this collection, more than 90% of the animations are produced by individual animators who attended animation workshops by Goethe Institut and Afghan-Film Organization; however, there are some other animations which are supported financially by Goethe Institut and some other animations that are made independently by Afghan animators and have been projected in "AfghAnimation Animation Film Festival". Most of the animations created in Afghanistan are sponsored by "Goethe Institut, "Republique Francaise" and "British Council". The Animation department of Afghan Film did not last very long due to the lack of professionals and financial challenges. The department is now closed. Ms. Sahraa Karimi said this was despite the announcement of several positions for this department which were not filled.

Sahraa Karimi, who has a doctorate in Cinema (Fiction Film Directing and Screening) from the Film and Television Faculty of the Academy of Music and Performing Arts in Bratislava, Slovakia, directed 30 short films, three documentaries and a 90-minute fiction film. She was the first ever female head of Afghan Film. She fled Afghanistan in August 2021 when the Taliban returned to power. Since then, the future of the Afghan Film organization has remained uncertain. Due to security challenges, the animation film festivals were no longer held.

Afghan Animators and their Animations

This list is organized based on Afghan Film-animation archive (the collection of 61 animations in 4 DVDs) and from online sources. The list is arranged based on the animation techniques used.

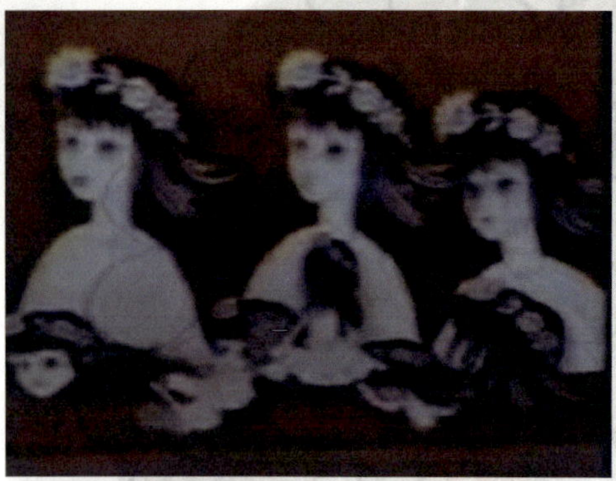

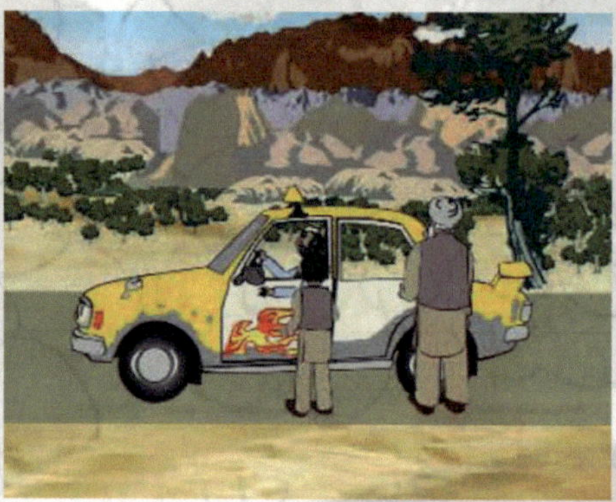

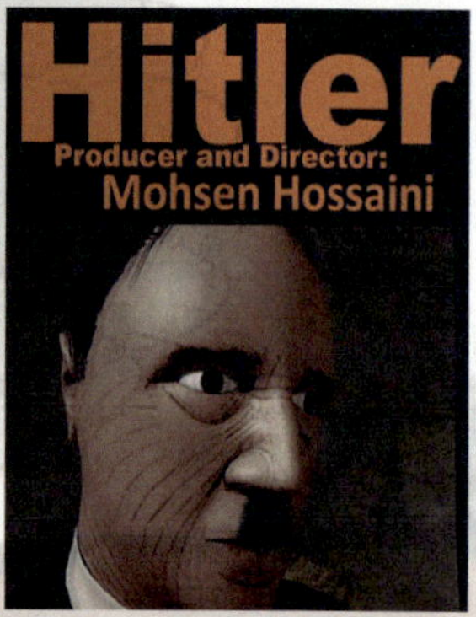

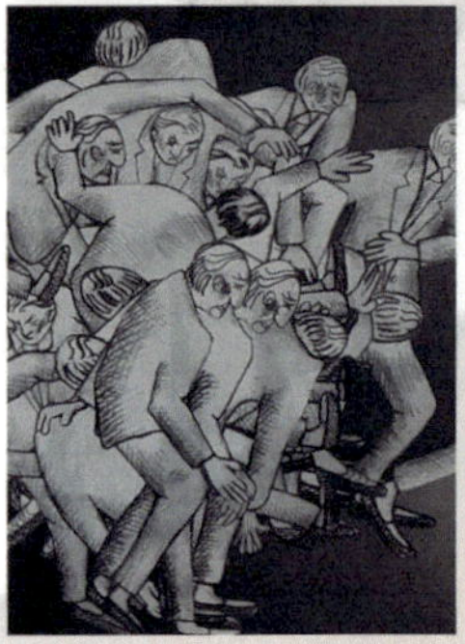

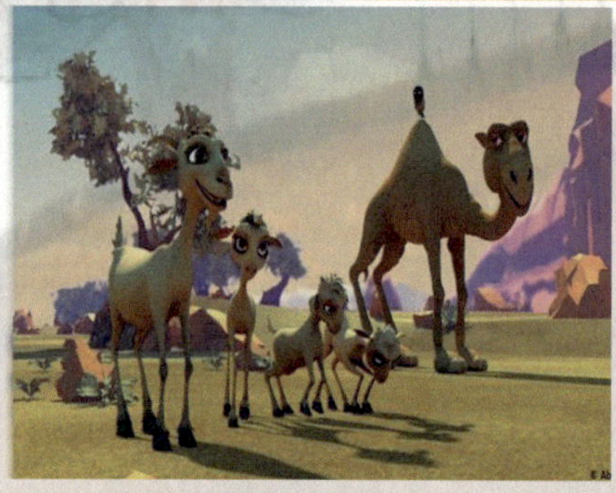

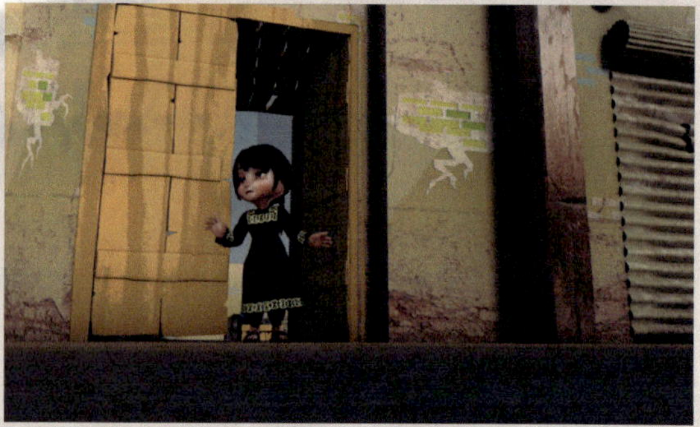

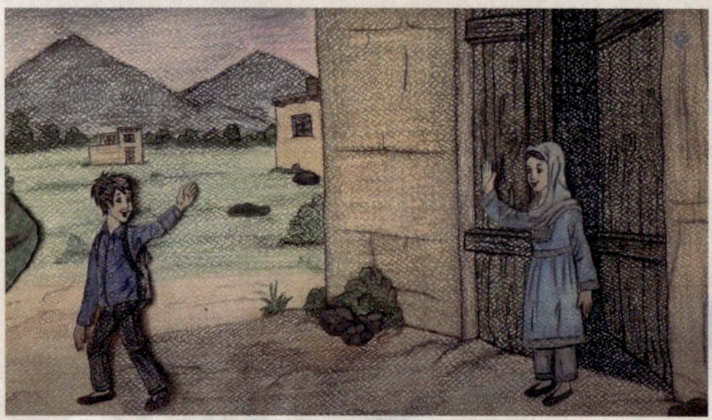

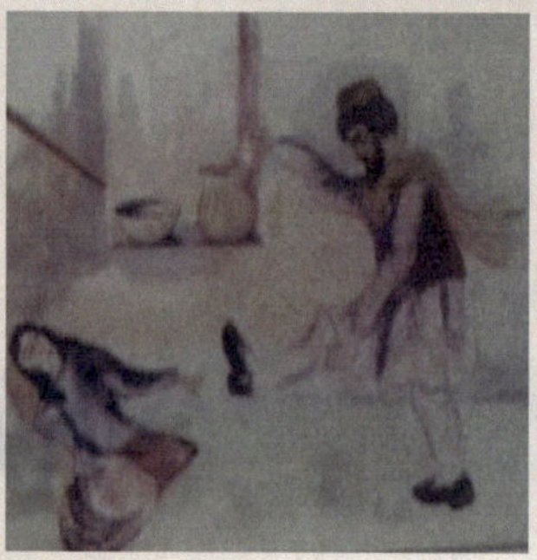

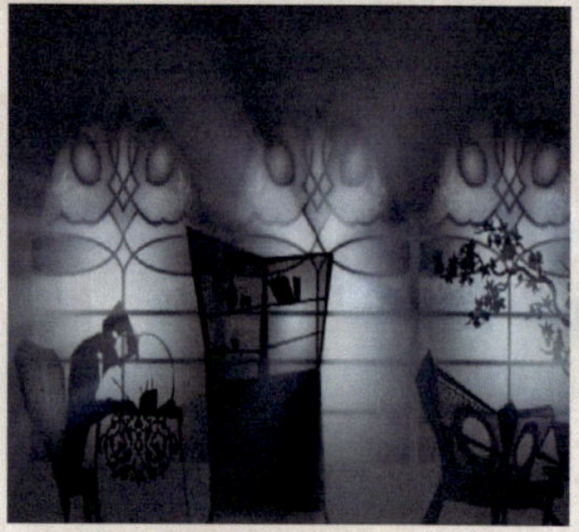

Table1: List of Afghan animators and their animations (Traditional Animation)

No	TITLE	ANIMATOR	TYPE	DURATION MIN:SEC	YEAR
1	Shelter	Mohsen Hossaini	Traditional	5:47	2006
2	Hitler	Mohsen Hossaini	Traditional	6:42	2009
3	Perfume	Mohsen Hossaini	Traditional		2009
4	Event	Mohsen Hossaini	Traditional	5:34	2019
5	The Reel	Lajaward Haqiqi	Traditional	3:30	2012
6	Schizophrenia	Lajaward Haqiqi	Traditional	4:30	2013
7	Play on the play	Lajaward Haqiqi	Traditional	4:00	2013
8	Remorse	Lajaward Haqiqi	Traditional	4:00	2011
9	Black and White	Lajaward Haqiqi	Traditional	3:00	2010
10	Rescue	Lajaward Haqiqi	Traditional	2:50	2014
11	Red Apple	Lajaward Haqiqi	Traditional	2:50	2015
12	Please Flush Down	Narges Noori, Fatema Hasani and Gholamreza Majidi	Traditional	3:15	
13	Softly and continuously	Ehsanullah Wakili	Traditional	2:00	
14	Cacao	Moh. Rafi Bayan	Traditional	2:00	
15	Clock	Mohammad Akrami	Traditional	1:40	2015
16	Rose to Mother	Faiza Amini	Traditional	1:20	2015
17	Once Again wish	Shamim Haqshenas	Traditional	2:00	2015
18	Don't Judge Me	Freshta Safi	Traditional	1:20	2015
19	The Bird and The Ant	Ehsan Ahmadi	Traditional	1:45	
20	Love Compensation	Amaina Mosawi and Nazgol Rafiye	Traditional	3:30	
21	Cell Phone Addiction	Faisal Zahib	Traditional	1:30	2015
22	Part of My Life	Hasibullah Sediqi	Traditional	1. 2:30	
23	I wish	Mahdi Arman	Traditional	1:00	
24	Silent Shout	Abulfazl Warasta	Traditional	20seconds	

Table 2: List of Afghan animators and their animations (Stop motion + Experimental Animation)

No	TITLE	ANIMATOR	TYPE	DURATION	YEAR
1	A Latter to Paradise	Moh. Nasir Hashimi,	Stop motion	6:00	
2	Death of Freedom	Moh. Nasir Hashimi	Stop motion	3:20	2010
3	Kabul in 2014	Moh. Nasir Hashimi	Stop motion	2:40	2013
4	Station	Yega Moqarabi	Stop motion	2:15	
5	The Destroyed Civilization	Fatema Hassani	Stop motion	1:30	2013
6	Governor	Meahdi Arman	Stop motion	2:30	2011
7	Cigarette	Hizbullah Alami	Stop motion	1:50	
8	Playing with Numbers	Wajiha Sayar	Stop motion	3:20	2011
9	The Savior	Mahdi poya	Stop motion	2:15	
10	The Balance	Naseem Khogyani	Stop motion	2:00	2011
11	Concomitant	Tahira Jamshidi	Stop motion	1:00	
12	My Dream	Mansoreh Ansari	Stop motion	3:00	
13	SH Promotion	Mursal Azizi	Stop motion	1:30	2019
14	Solidarity	Mursal Azizi	Stop motion	2:43	2021
15	Dream of Peace	Mursal Azizi	Stop motion	1:00	2021

Table 3: List of Afghan animators and their animations (2D Animation)

NO	TITLE	ANIMATOR	TYPE	DURATION	YEAR
1	Begin Again	Mohammad Daud Hedayati		2:38	2014
2	The Election	Mohammad Hussain Panahi	2D	1:00	
3	The Dream	Mohammad Hussain Panahi	2D	1:00	
4	The Tiny Dream	Mohsen Shams	2D	3:40	2013
5	The Greedy	Mahdi Bakhtiari	2D	2:20	2013
6	Life Imprisonment	Gholamreza Majidi	2D	3:30	2013
7	Terrorist	Sayed Nazim Khelwaty	2D	50 seconds	2010
8	Truck	Moh. Rafi Baian	2D	3:00	2011
9	Jealousy	Rafi Bayan	2D	2:40	2011
10	Wish	Fatema Hassani	2D	1:50	2011
11	Think Different	Mohamad Ansari	2D	1:00	
12	Street Harassment		2D	1:00	
13	Saved	Mehdi Bakhtiari	2D	2:39	
14	The Life	Mohammad Reza Nawruzi	2D	6:50	2011
15	White	Sakina Mohammadi	2D	3:00	
16	Hunt	Rana Ahmadi	2D	1:50	
17	Fifty Fifty	Mari Ahmadi	2D	3:00	
18	The Last Page	Mohammad Nasir Hashimi	2D	3:00	2011
19	Black balloon	Alireza Haidary	2D		2017
20	Funny and Kitty	Mursal Azizi	2D	1:30	2018

Table 4: List of Afghan animators and their animations (3D Animations)

No	TITLE	ANIMATOR	TYPE	DURRATION	YEAR
1	Afghan Doll	Mohammad Aref Abedian	3D	4:25	
2	One Wish	Mohammad Aref Abedian	3D	3:15	2013
3	Life Is Going On	Mohammad Aref Abedian	3D	30 seconds	
4	Monster	Mohammad Aref Abedian	3D	2:10	
5	P.H	Hesam Pirqadam	3D	5:00	2013
6	Childhood World	Hesam Perqadam	3D	2:15	
7	Obscenities	Mohammad Daud Hedayati	3D	5:44	2012
8	Tree	Fatema Hassani	3D	50 seconds	
9	Window	Fatema Hasani	3D	1:30	
10	Maadar	Sayed Alireza Sajadi	3D	3:00	
11	Addiction	Nori	3D	2:00	
12	Gel	Wasey Rahimy	3D	2:30	2013
13	Facebook Bazi	Hafiz Afzail	3D	1:58	
14	Mountain girl	Hafiz Afzali	3D	6:52	
15	The Dark Time	Hussain Panahi	3D		
16	White Line	Ali Reza Sajjadi	3D	1:51	
17	Yassin and kaka Rauf animation series	FKH Media for USID	2D 3D	8 episodes	2005-2009
18	Buz-e-Chini	Abas Ali Wahdat	3D	20 minutes	2011

Types of Animation Techniques Used in Afghanistan

There are several different types of animation that have been used in Afghanistan. Some common types include:

1. Traditional Animation
2. 2D Animation
3. 3D Animation
4. Stop motion
5. Sand Animation
6. Experimental Animation

Based on archival evidence, Afghan animators are familiar with all of the above types of animation techniques. This list has introduced 75 animations, among these animations there are 18 in 3D, 20 in 2D, 15 using Stop motion and 24 are made in traditional format. Buz-e-Chini is the longest animation made in the history of animation in Afghanistan (20 minutes) and is directed by Abas Ali Wahdat. The first female animator in Afghanistan is Lajaward Haqiqi. She made her first animation "Black and White"in 2010. Mohammad Aref Abedian and Mohammad Daud Hedayati are recognized as some of the best 3D animation directors, details are given below.

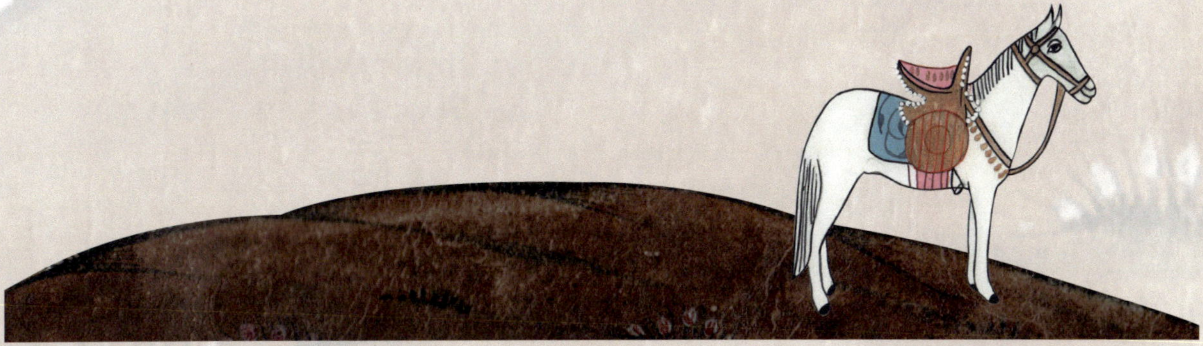

The First Animation movie, Buz-e-Chini

Buz-e-Chini is the longest computer-generated 3D film in Afghanistan. The film was released on September 24, 2011 with a duration of 20 minutes. The film was directed by Abbas Ali Wahdat who was born in Afghanistan but fled the country when the Taliban took over in 1996. He studied animation in Pakistan. The film has been produced mostly in Pakistan by Post Amazers, based in Karachi.

The story is based on a folklore tale of the Hazara people of Hazarajat in the central highlands of Afghanistan, where a cunning wolf tries to cheat three kids of a goat in the valley of Bamiyan. This film was dedicated to Hussain Ali Yousafi, who played the role of Baiyo (the wolf) in the movie and who was assassinated in Quetta, Pakistan before the film was released.

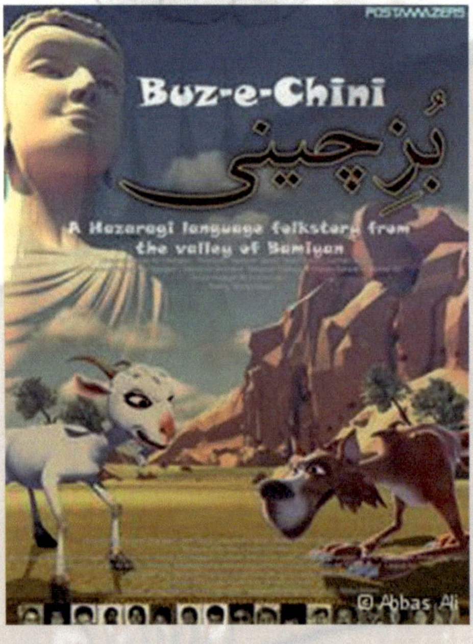

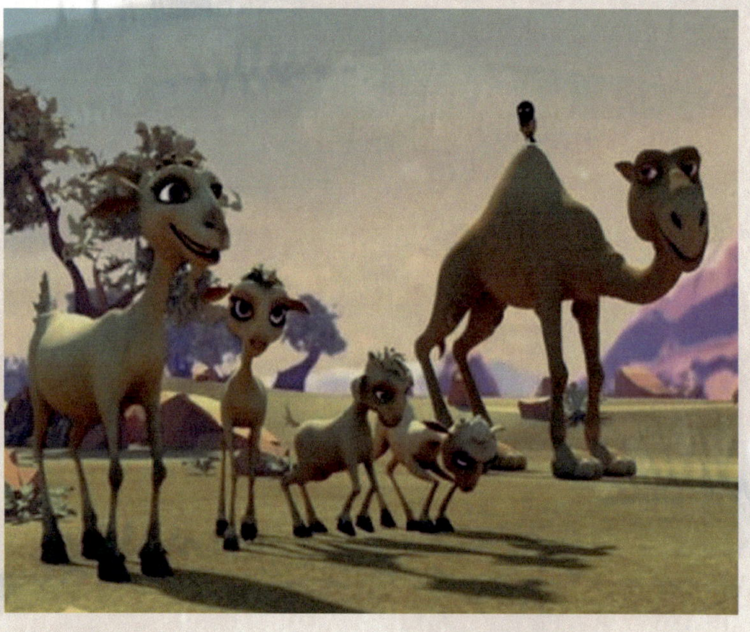

First Female Animator, Lajaward Haqiqi

Lajaward Haqiqi was born in 1990 in Kabul. In 2011, she completed a Bachelor's degree in Graphic Design from Kabul University. Since then, she also studied animation in Goethe Institut. She has made seven short-animated films which have been screened in national and international festivals. One of her animations, "The Reel" won the 'Best Animation Film' award at the Eunic Film Festival, which was held in Kabul in 2012. The animation is about 03:15 minutes and depicts listening to music as a legitimate need for all human beings and its prohibition as an inhuman act. It reminds the people of Afghanistan of a recent past so that they can compare the past deprivation of freedom with the present achievements.

She showcased two of her animated films, "The Reel" and "The Play on Play" at the 11th Asian Women Film Festival, organized by the India chapter of the International Association of Women in Radio and Television (IAWRT) in New Delhi. The length of each animation is around 4:00 minutes. The event showcased an exciting selection of 60 films and sound works from a range of countries including Afghanistan, Bangladesh, Belgium, China, Germany, Hong Kong, Japan, India, Iran, Israel, Switzerland, Taiwan, the UK and the USA.

In 2013, her animation was screened at the AfghAnimation Film Festival in Kabul. Her animation "The Reel" has also been screened at the SAMA International Film Festival of Sweden. This Film Festival aims to develop and support creative cinema in Afghanistan and display more Afghan filmmakers as well as some foreign filmmakers that have made films regarding Afghanistan. Later, her animation "Play on the Play" was showcased in the second Afghanistan Human Rights Film Festival in which 46 films among 380 films were selected from national and international films. Lajaward's animation selected as one of the 23 best films in this festival.

She has also participated in some other festivals and has collected several certificates and awards. Currently, Lajaward Haqiqi has relocated to France. She prefers to use mixed media to create artistic pieces.

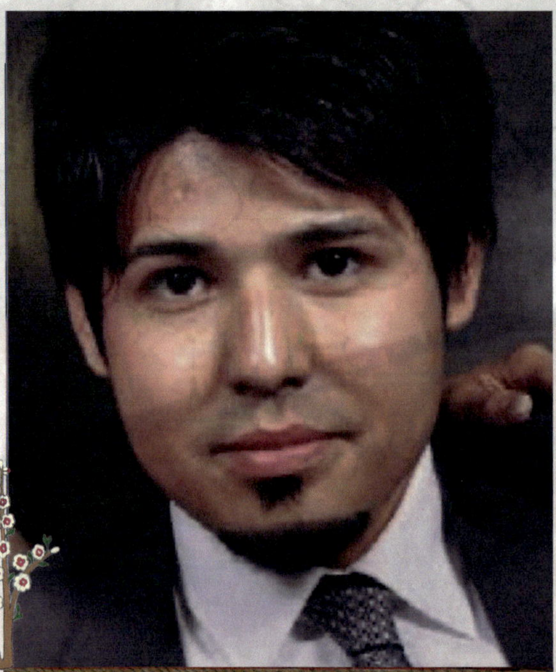

Mohammad Aref Abedian, Animator

Mohammad Aref Abedian, Animator

Mohammad Abedian was born in Kabul in 1987. He completed his Bachelo's degree in Multimedia from the private institute "Arna" when he was a refugee in Iran. Later, he completed a 3D animation course and advanced his professional career in Didar Animation Production Company and other animation companies. In the fall of 2008, he returned to Afghanistan and began working in Afghan medias. So far, he has directed four animation films namely, "Monster", "A Wish", "Cage" and "Afghan Doll".

He was awarded the first prize in first AfghAnimation Animation Film Festival for the short 3D animated film "A Wish", which is around four minutes and was made in 2011. In this film, a dandelion acts as a messenger for a little child in war-ravaged Afghanistan. The dandelion travels around and views all the bitter facts until it reaches a little girl. The dandelion tries to express the dream of all Afghan children, living in peace and harmony with their parents, a dream that has been denied to them for decades. Mr. Abedian left Afghanistan and is currently living in Germany.

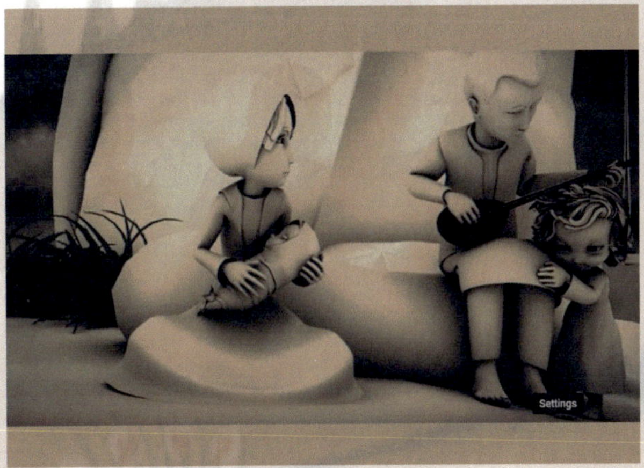

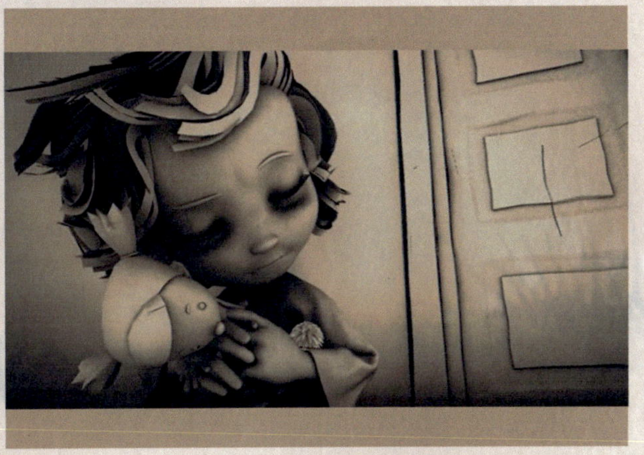

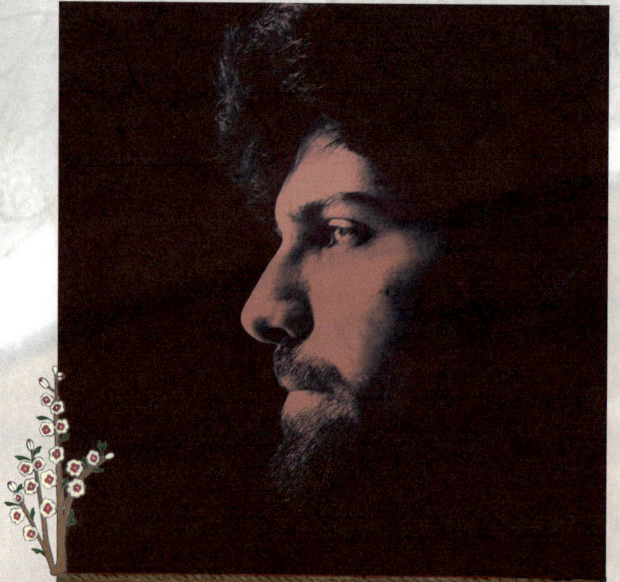

Mohammad Daud Hedayati, Animator

Mohammad Daud Hedayati, Animator

Mohammad Daud Hedayati was born in 1990. He was interested in arts from his childhood. In 2008 he graduated from the GMMC Art Center. Shortly after he obtained an additional degree in Graphic design from the ECC (Korean Institute) in Kabul.

He is interested in most art forms and has also been working in the visual arts since 2012, creating 2D & 3D animations and visual effects, taking these through from pre-production to post-production. He has made two short animated films titled "Obscenities" and "Begin Again". He is keen to pass on his knowledge to the benefit of fellow artists which is why he, from time to time, has organized free workshops in Arts.

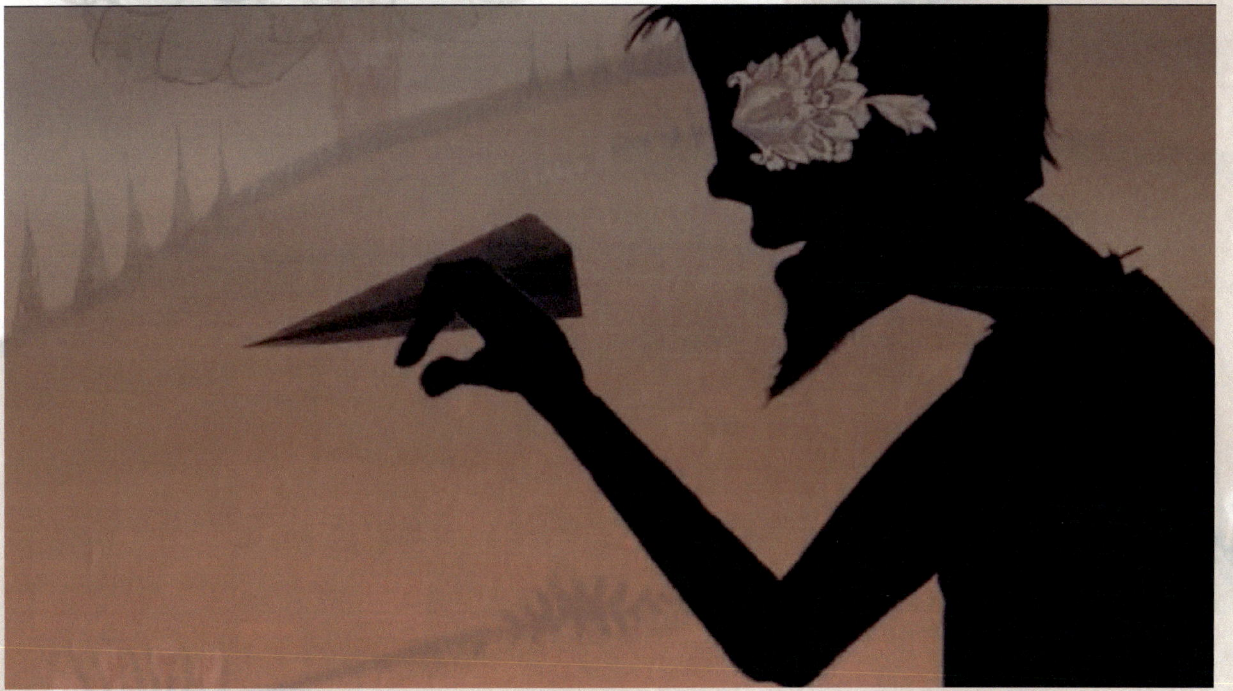

Since graduating, he has exhibited his work in various exhibitions across Afghanistan as well as abroad. This includes:

- March 2018 contributing artist at the "Women's days art festival", Afghan Cultural house and UN. Kabul, Afghanistan.
- July 2015 contributing artist at the "Kabul Artists", Afghanistan Fine Art Faculty, Kabul.
- June 2014 contributing artist at the " Nimm Wahr (Perceive)"-Exhibition, Goethe Institute Kabul, Afghanistan.
- 2013 & 2014 Afghan Animation Festival at the Goethe Institut , Kabul, Afghanistan.
- December 2014 "The Impact" an exhibition at the Goethe Institute, Kabul.
- November 2013 contributing artist "Art Prize Competition" at Bagh-e Babur: Queen's Palace, Kabul, Afghanistan.
- July/August 2012 contributing artist for the "Royal Academy Summer Exhibition" London, UK
- December 2010 contributing artist for an exhibition at the French Cultural Centre, Kabul, Afghanistan.
- 2008 contributing artist at the GMMC-exhibition, Kabul, Afghanistan.

His awards include:

- 2018 winner of "Women's day live drawing competition" conducted by Art Business Consultants, Kabul, Afghanistan.
- 2013 winner of "Afghanistan Animation Festival" conducted by the Goethe Institute in Afghanistan.
- 2012 winner of "Afghanistan Contemporary Art Price Competition" conducted by Turquoise Mountain, Kabul, Afghanistan.
- 2011 winner of "Afghanistan Social Enterprise" challenge conducted by the British Council of Afghanistan

Sara Barackzay, Afghan Artist

Sara Barack is considered to be the first female Afghan artist who has completed a Bachelor's degree in Animation. She graduated from Baskent Universitesi, Turkey, in 2019 but has not produced any animated films as yet.

Despite facing significant challenges there are still a number of young and talented Afghan artists (both female and male) who are exploring different animation techniques for a range of content including commercials, educational videos, video clips for songs and comedy etc.

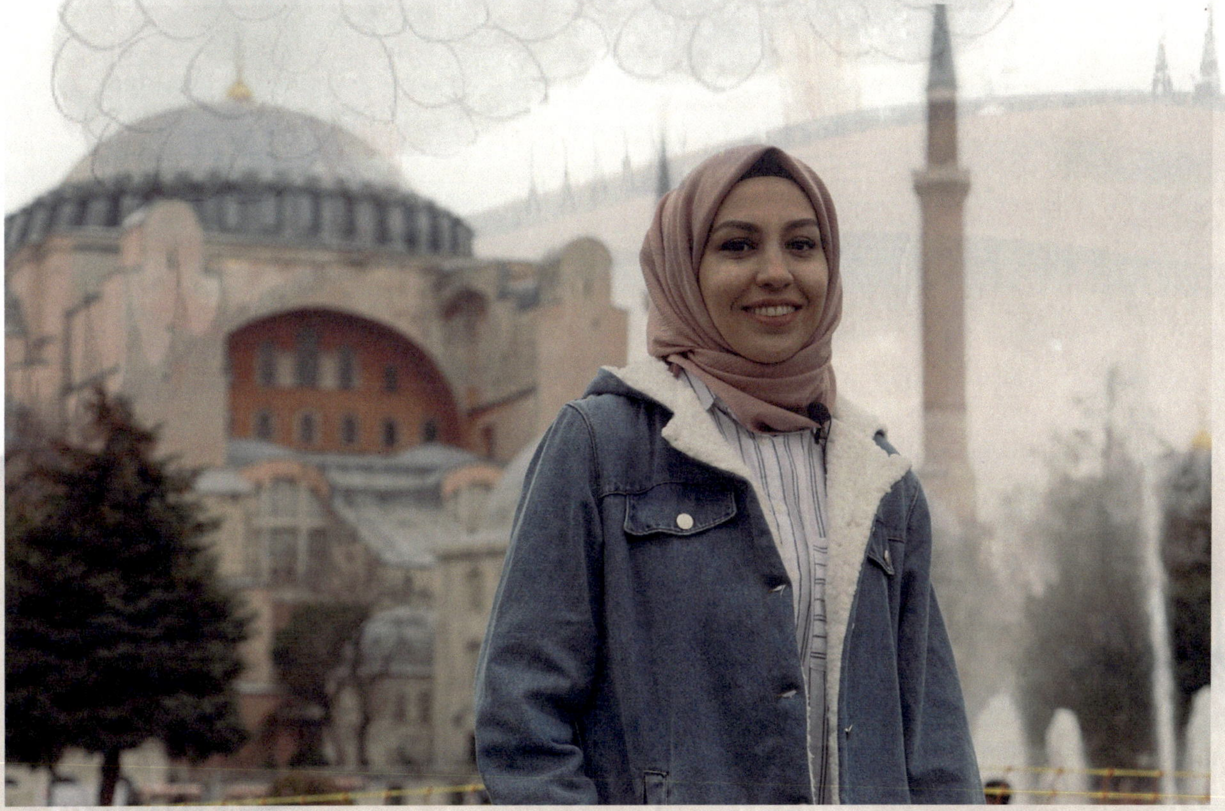

Zubair Ahmad Kakarr, Afghan puppeteer

Zubair Ahmad Kakarr is an Afghan puppeteer who has worked as a TV producer, production assistant, voice talent, writer for live action films (LAF) and puppeteer in Tolo and Lemar TV, from 2011 to 2017. During his work at this TV channels, he attended workshops including puppet training and writing in India and the UAE.

In 2019 he made his own puppet "Fazayee", a special Afghan puppet for children, and joined Nation Radio Television of Afghanistan (RTA) as director, puppeteer and Youtube editor.

Fazayee conducted inspiring interviews with various celebrities, including:

- Cricket stars Rashid Khan, Mohammad Nabi, and Hashmat, sharing their insights on the sport
- Musician Shafiq Mureed, discussing about Afghan music
- Members of the Zohra Orchestra and students from the Music Institute of Afghanistan (Music School), highlighting the importance of music education
- Media and art personalities, sharing their stories and experiences to inspire the young generation

Zubair tries to create new puppets and produce interesting educational and entertainment programs with his own puppet creations to bring positive changes to society. He plans to make a specific children's TV show for RTA. His dream is to open a local digital TV station and organisation for Afghan children where the children can come and meet their favourite puppet characters, play with them to experiment and be educated.

In 2020, Zubair created and performed a COVID-19 awareness campaign using a walking Muppet character, educating people about the pandemic. He wrote and acted in short videos featuring the COVID-19 character, collaborating with a talented young team at RTA. Zubair envisions a children's TV show for RTA and dreams of establishing a local digital TV station and organization where Afghan children can interact with their favorite puppet characters, learning through play and experimentation.

To achieve this, Zubair founded and registered Child AFG (طفل افغان), an online education center for children aged 3-9, with the Ministry of Education. As the CEO, he aims to provide a platform for future generations to learn through entertainment. Child AFG offers a unique space for kids to engage in fun, interactive educational experiences.

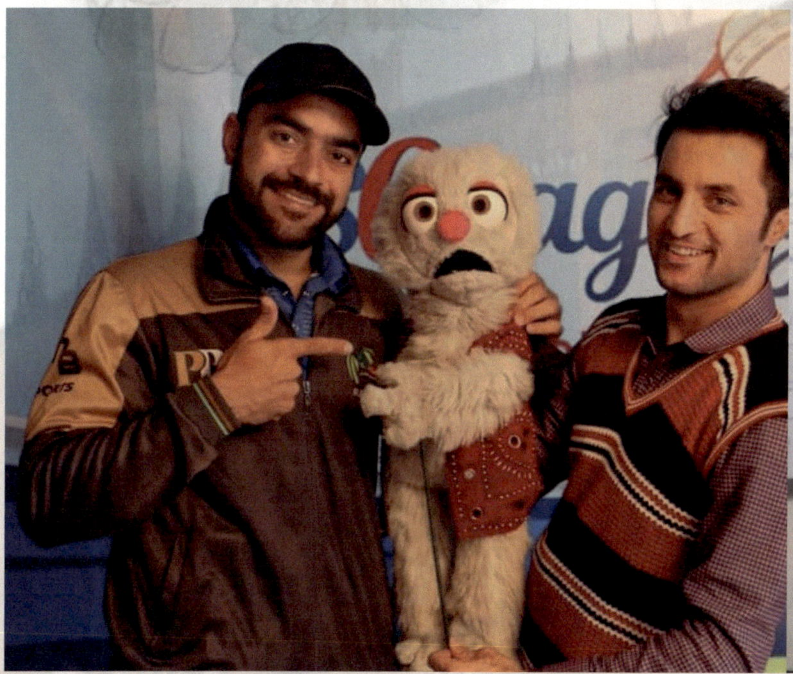

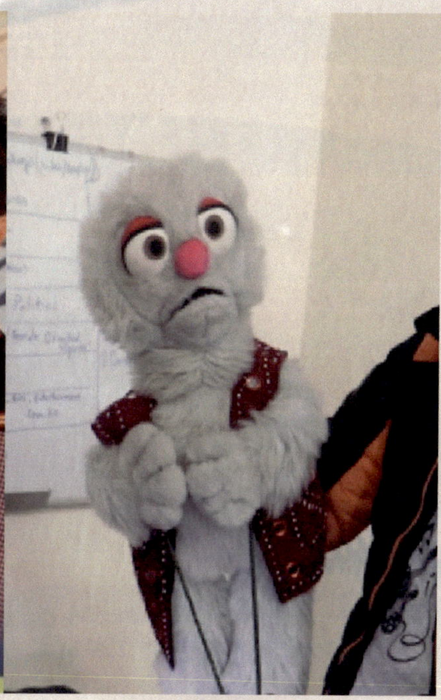

Baghch-e-Simsim, Children's Television Series

"Baghch-e-Simsim" (Sesame Garden) is the Afghan adaptation of Sesame Street, produced in collaboration with Tolo TV and Sesame Workshop. Launched in 2011 in Afghanistan, it features beloved characters like Elmo and Big Bird, with the aim of enhancing education for Afghan children.

The show combines Muppets and educational shorts, but some characters were omitted due to cultural sensitivities. This Afghan version is part of a broader global effort, with similar productions in Pakistan, Bangladesh, Egypt, Mexico, and Russia. "Baghch-e-Simsim" has brought joy and learning to Afghan children in both Farsi/Dari and Pashto languages the local languages of Afghanistan, with the introduction of local Muppet character Zari, a curious and kind 6-year-old girl, and her younger brother Zeerak, who promotes gender equality and respect. The show focuses on literacy, math, and life skills, with an emphasis on girls' education, mutual respect, diversity, and national identity, reaching millions of Afghan children.

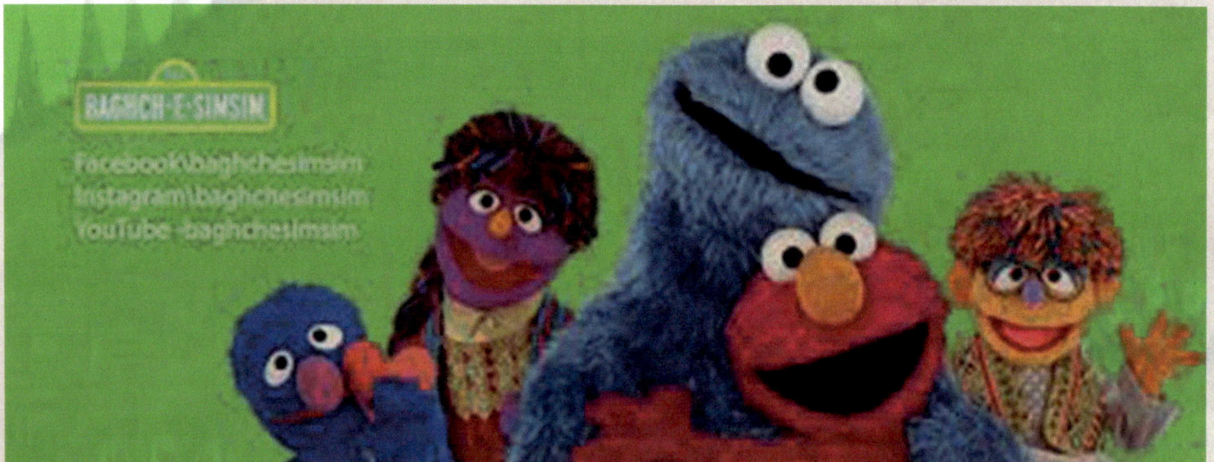

The Animation Studio, Kapila Motion Pictures (KMP)

From my interview with Mr. Mohammad Nasir Hashimi, Director of Kapila Motion Pictures in January 2020 in Kabul, I found that Kapila Motion Pictures KMP began operating in 2018 with a small team of 13 members. Nasir and his team in Kapila Motion Pictures worked on an animation series called "Yusuf and Zuleikha", produced by Pascale Bastide and directed by Patrick Pleutin and Nasir Hashimi with support of the French Institute of Afghanistan and the US Embassy in Kabul.

The story of this series is based on the "Mawlana Jami's Poems" from 15th century and the characters and environments are designed based on old traditional miniatures collected from the National Museum of Afghanistan. The plan was to make 59 episodes, each episode with a length of two to three minutes. By 2020, 7 episodes were made using eight frames per second in traditional form. Most of the animators working in KMP were trained by the Goethe Institut incuding Naser Hashemi and Lajaward Haqiqi.

I conducted an online interview with Pascale Bastide, the producer of the "Yusuf and Zuleikha" animated series, and she said the project's objective was to make Afghan culture alive and that there was no one better than Afghan artists to do so. The agreement they had with Masuma Nazari, Director of National Archives Museum at the time, was to use images from the Museum. As such, they chose a series of miniatures telling the story of Yusuf & Zuleikha. To complete this, they used Jami poems which were edited by a scholar.

Bastide said "We had decided that the format would be 3mn film for each miniature. After meeting different artists, I chose to work with Nasir Hashimi who was head of Kapila productions. We decided to produce 7 episodes so it was 20 mn of animation that had to be produced. I spent almost 2 years with the team, who had never produced such a length of animation, so while producing we had to teach them organization, rigor, methodology etc.

Kapila produced all the 2D animations, design and coloring, the composition was made in France as well as sound effects and mixing. 12 people worked with Kapila; Nasir was co-director with Patrick Pleutin.

In conclusion at the end of the production I have to say that Kapila team was at the level of international standards in animation."

Nasir Hashimi was born in 1990 in Baghlan province. He completed a Bachelor of Cinema in the Fine Arts Faculty of Kabul University and participated in animation workshops conducted by Goethe Institut. Hashemi has directed 3 short animated films, namely "Death to Freedom", "Kabul in 2014" and "A Letter to Paradise". Moreover, he has also directed three short films.

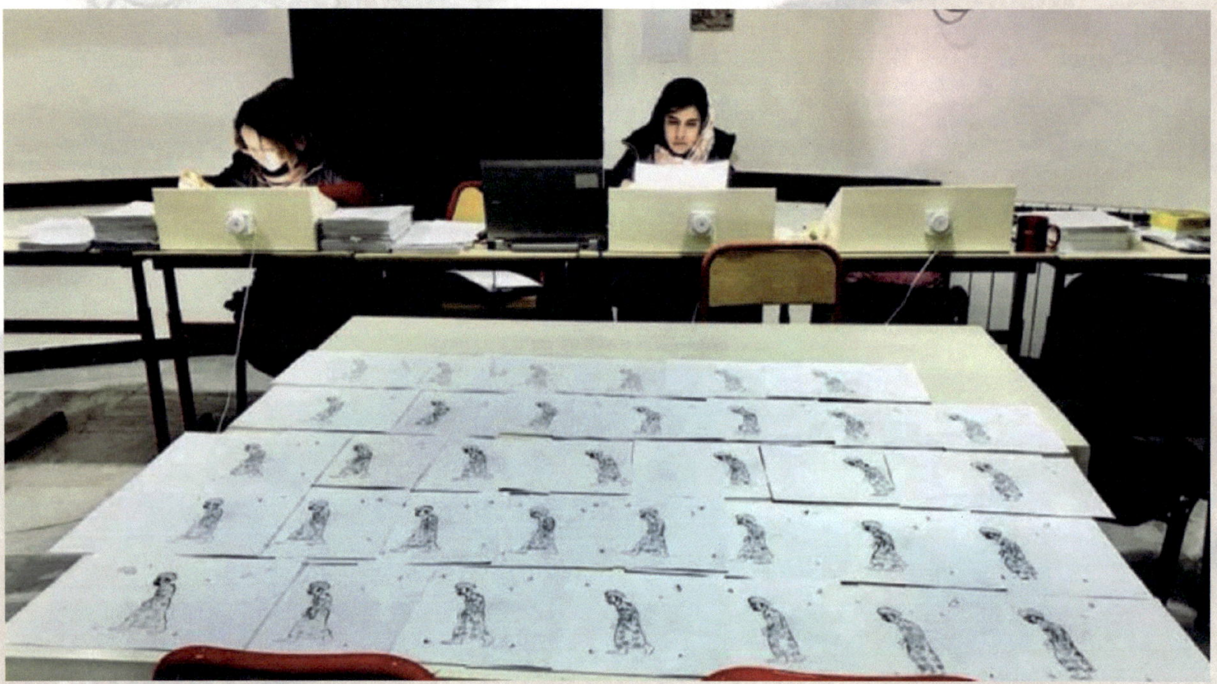

He has worked as Head of the Animation Department of Afghan Film Organization and has played an important role in organizing animation festivals and animation workshops supported by Goethe Institut. He has collected almost all the animations produced in Afghanistan which are now available in the Afghan Film archive.

He has also worked as an assistant director, production manager and producer in both national and international films and has won prizes in different national and international film festivals. He pursued his Masters in Film Production at the Beijing Film Academy in China. Nasir Hashimi, along with some of their team members, is now in France trying to create paths to continue their work. They have worked on two new projects recently - "Per-e-Yakhsoz" and "Azhdaha" – which are based on the old takes from Afghanistan.

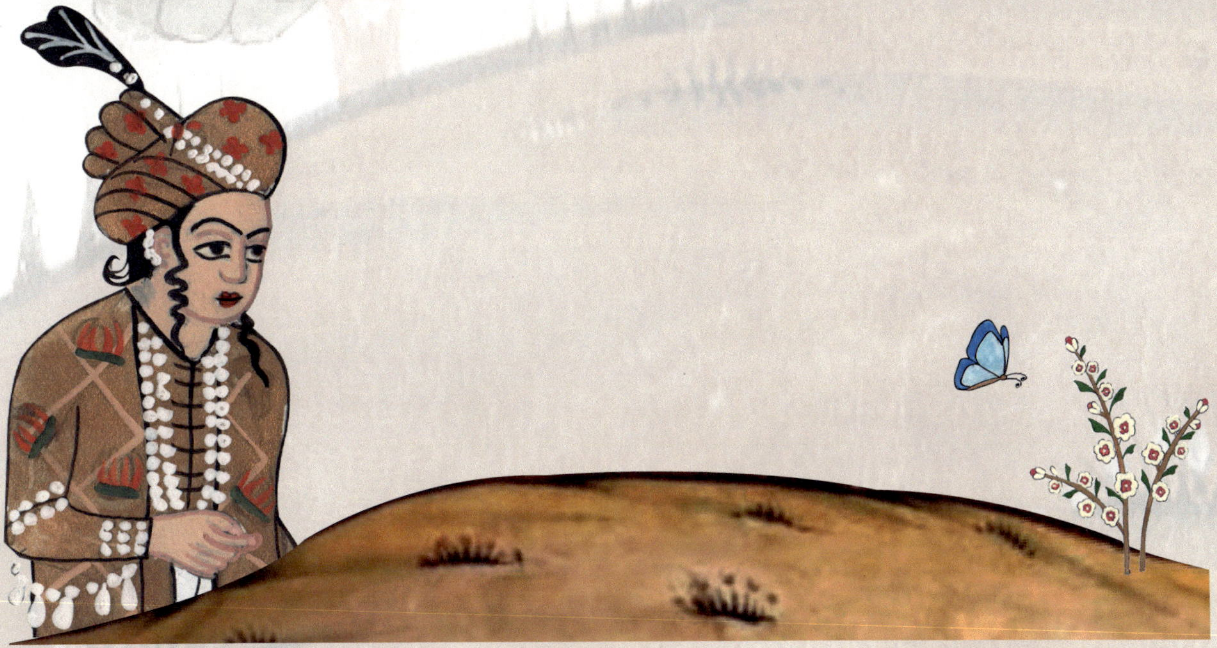

Zawia Media (Production Company)

Zawia is a digital media house located in Kabul, Afghanistan. It was launched in January 2019 with the aim of emphasizing the untold realities of Afghanistan. Unlike many other media houses that try to show daily conflicts, Zawia not only strives to inform and update citizens, but also aims to inspire, motivate, and empower the new generation of Afghan citizens through portraying a different, beautiful image of the country.

During my interview with Mr. Mujtaba Karimi, production director of Zawia Media on 8 January 2020, in Kabul, I found that Zawia Media is a platform for social media users working in three sections; news, video production and animation.

The animation section comes under "Zawia Lolz" in which only short animated film are displayed once a week through social media for social awareness. A team of eight animators (trained through online courses) produces these 2D animated short films with a duration of two minutes each. This department uses a specific pipeline and the software called "Anime Studio" for making these animations.

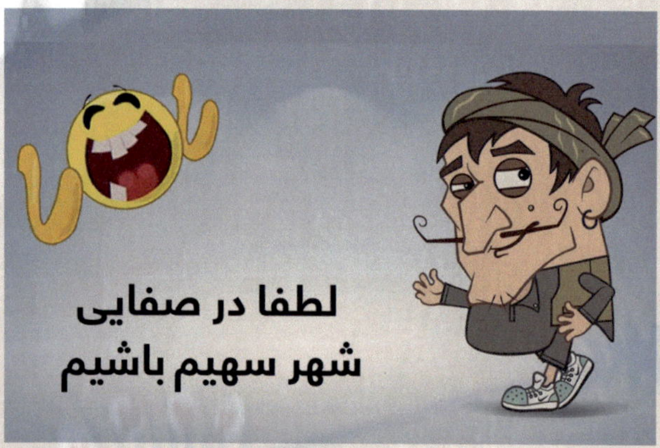

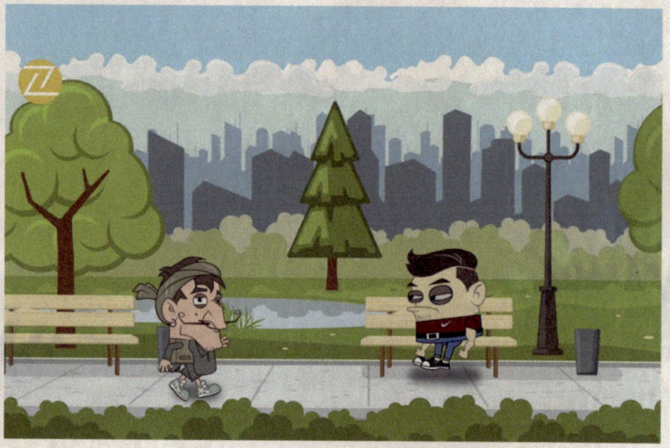

Festivals and awards

Film festivals have played a vital role in the introduction and growth of cinema and animation. In order to get more information about Afghan animations, I reviewed 16 brochures of different national film festivals. Names and titles of these brochures are as follows:

1. The AfghAnimation Animation Film Festival, 2013;
2. The Second AfghAnimation Animation Film Festival, 2014;
3. The First Afghan Human Rights Film Festival, 2011;
4. The Second Afghan Human Right Film Festival, 2013;
5. The Third Afghan Human Rights Film Festival, December 2015;
6. First Police Film Festival, May 2012;
7. The Eunic Film Festival, September 2012;
8. First Herat International Women's Film Festival, 2013
9. Second Herat International Women's Film Festival, 2014
10. Third Herat International Women's Film Festival, 2015
11. Fourth Herat International Women's Film Festival, 2016
12. Mehrgan International Film Festival, 2013
13. Mehrgan International Film Festival, 2014
14. Mehrgan International Film Festival, 2015
15. Mehrgan International Film Festival, 2016
16. Mehrgan International Film Festival, 2017

After reviewing these brochures, I found that most of the animations have been screened in these festivals and the animators have received certificates and awards from these festivals.

Here, I introduce two of the popular festivals in Afghanistan which support Afghan filmmakers in the country.

Herat International Women's Film Festival

The Herat International Women's Film Festival (HIWFF) is an iconic and the most prestigious Women's Film Festival in the region. It was established by Roya Film House and Armanshahr Foundation/OPEN ASIA in 2013. It is an annual event held in Afghanistan; gathering filmmakers, cinema lovers, and activists from all over the world. HIWFF tackles issues related to women rights, gender equality, sexual identity, domestic violence among others through the power of the lens and cinema which is the window to society and in this case particularly women.

The first festival was held in 2013 in the ancient Alexander's Citadel of Herat for three days, attracting more than 100 films from leading filmmakers from 20 countries. More than 2,000 people watched the films. The second and third rounds of the festival were also held in the ancient Citadel of Herat, while the fourth round was held at the Garden of Babur (Bagh-e-Babur) in Kabul. The fifth edition was held at the Intercontinental Hotel in Kabul and attracted 1000 films from the region and across the world.

Many renowned national and international filmmakers have attended the various rounds of the festival, numerous media outlets covered the festival, and thousands of people have watched the film selections. The Herat International Women's Film Festival is perhaps the most important cultural event of its kind in the Heart of Asia. Several awards have been offered including best feature film, best documentary, best short film and best animation.

Animation has featured in various editions of the festival. At the first round of festival seven animations were selected for competition. In the second round of the festival, five animations were selected for competition. A year later, in the third round, seven animations were selected, while in fourth round which was held in 2016, six animations were selected. There were no animations featured in the fifth round of the festival in 2019. Since the fall of Afghanistan under the Taliban regime in 2021, the festival has not been held in Afghanistan.

Mehrgan International Film Festival

I interviewed Mr. Ali Mohammadi Kiyan, Director of Mehrgan International Film Festival in January 2020, in Kabul. In addition to the interview, I reviewed brochures for each edition of the festival. Mehrgan International Film Festival is a valuable practice for the development of culture and art in Afghanistan, and an opportunity to discover new talents in the country's cinema and film making area.

The secretariat of the festival, together with the artistic conferences of the Academy of Art and other partners, held five rounds of the Mehrgan film festivals from 2013 to 2019 with a view to improve public opinions, increase the visual literacy of Afghan society, and provide a platform for making valuable artwork. As an important artistic event, these festivals have played an important role in encouraging young filmmakers and creating a sense of unity among citizens.

The sixth round of this festival took place in November 20, 2020. Mr. Ali Mohammadi Kiyan is the president of Academy of Arts and Cinematic Sciences. The Mehrgan International Film Festival is part this Academy.

In every Mehrgan Film Festival, there are Selection Committees and Jury Committees. Both national and international films are shown in different categories including Documentary Films, Drama Films and Animation Films.

The first Mehrgan International Film Festival was held in January 2014, in the main hall of Afghanistan French Cultural Centre, Kabul. In this festival, there were two films in the animation section. In the national category, the name of the film was "A Wish". The length of the film was four minutes and was directed by Mohammad Abedian. In the international category, the name of the film was "A Wish" (An Afghan Girl and Her Dreams), directed by Fatima Hassani.

The first Mehrgan Film Festival received widespread domestic and foreign media coverage. It was reported on radio, TV, social networks, magazines, and national and international newspapers. The BBC, VOA and Deutsche Welle broadcasted it widely.

From 147 films delivered to the festival, 110 films were selected., 25 films were chosen for competition and 11 films for the guest division by the selecting committee. "A Wish" by Mohammad Aref Abedian, was part of selected films for competition.

The second Mehrgan International Film Festival was held in December 2014 in theMinistry of Education science Hall, Kabul. In the national section, there were no Animation films; however, in the international section, there were five films in Animation category; namely, "A Wedding Party" by Mahin Javaherian, "The Fox who Followed the Sound" by Fatemeh Goudarzi, "Bald Pigeon Fancier" by Sadegh Javadi, "Life Story of Red Pencil" by Ali Reza Chitaei and "Mammy Crow and the Fox" by Mohammad Ali Soleymanzadeh. In this program, 27 people were nominated for prizes and 11 received awards.

The third Mehrgan International Film Festival was held in September 2017 in the Russian Culture house, Afghan Film Hall and Salon of Art Academy, Kabul. In the national section, there were three Animation Films; namely, "Station" by Vega Moqarabi, "Clocks" by Mohammad Akrami and "Message (Manji)" by Mahdi Poyazadah. In the international section, there was only one Animation film by the name of "Junk Girl" by Mohammad Zare. In this festival, among 350 short films, 22 films were selected for the national competition, 22 were selected for international competition and 5 films were selected for special screening. The above-mentioned animations were part of 22 films selected for national competition.

The fourth Mehrgan International Film Festival was held in November 2018 in the Faculty of Fine Arts Hall, Afghan Film Hall and Art Academy Hall in Kabul. In the national section, there were no films in the Animation category; however, in the international section, there were two films in Animation category; namely, "Balance" by Barzan Rostami and "Empty View" by Ali Zare. In this festival, among 150 short films, 24 films were selected for the national competition and 23 films were selected for international competition. The fifth Mehrgan International Film Festival was held in 2019. In the national section, there was no film in Animation category; however, there were two animated films in international section; namely, "Inseparable" by Nicolas Blandin & Sabrina Blazquuez Y Gomez and "Safe Land" by Laleh Alipour. In this festival, 25 films were selected for the national competition and 21 films were selected for international competition.

Afghan stories featured in the international animation industry.

1. The Breadwinner

The Breadwinner (2017) is a powerful animated drama film directed by Nora Twomey and executive produced by Angelina Jolie. Based on Deborah Ellis' novel, the movie tells the story of Parvana, a young Afghan girl who disguises herself as a boy to support her family after her father's arrest by the Taliban.

With a talented voice cast, including Saara Chaudry as Parvana, and a powerful soundtrack featuring Afghan musicians and a haunting score by Jeff Danna and Mychael Danna, "The Breadwinner" has received widespread critical acclaim and numerous award nominations, including the Academy Award for Best Animated Feature.

This film has resonated globally, praised for its timely story, strong female lead, and beautiful animation, offering a powerful and emotional experience that highlights hope, resilience, and the human spirit's importance.

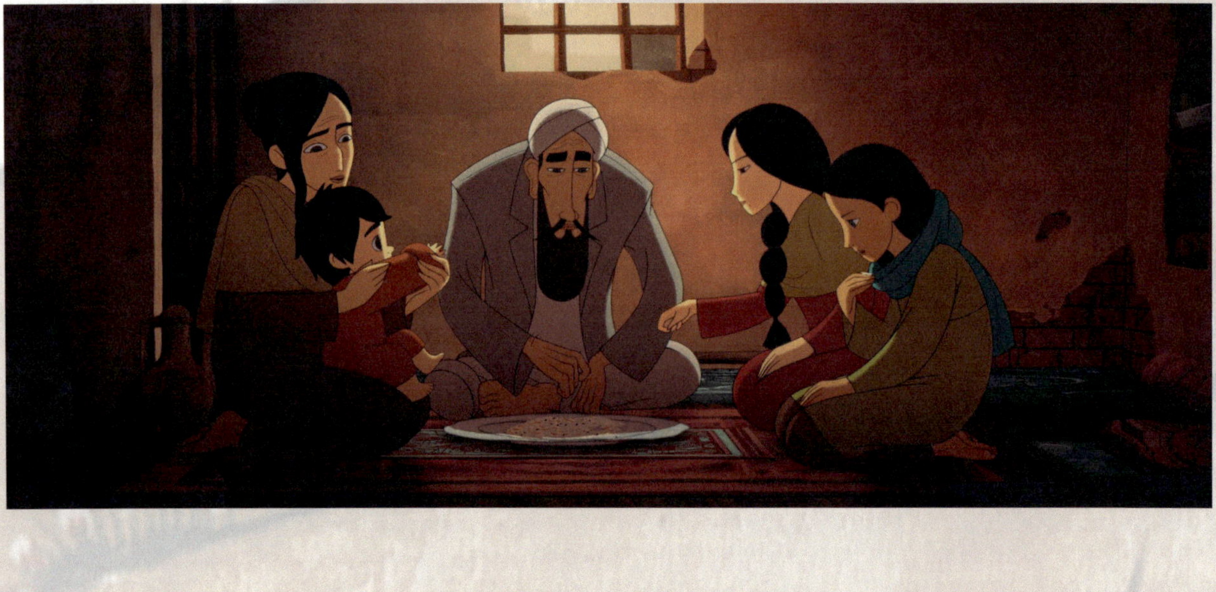

2. The Swallows of Kabul

The Swallows of Kabul is a novel by Yasmina Khadra, published in 2002, which was adapted into an animated film in 2019. The story takes place in Kabul, Afghanistan, during the Taliban regime and follows the lives of two couples, Mohsen and Zunaira, and Atiq and Musarrat, as they navigate the harsh realities of life under Taliban rule.

The animated film "The Swallows of Kabul" was a French-Luxembourgish coproduction, directed by Zabou Breitman and Éléa Gobbé-Mévellec. The film was produced by Les Armateurs (France) and Melusine Productions (Luxembourg), in association with Canal+ and France 3.

Through their stories, the novel and film highlight the brutal restrictions imposed by the Taliban, particularly on women, and the ways in which people find ways to resist and hold onto hope in the face of oppression.

The film received critical acclaim and won several awards, including the Annecy International Animation Festival's Grand Prix.

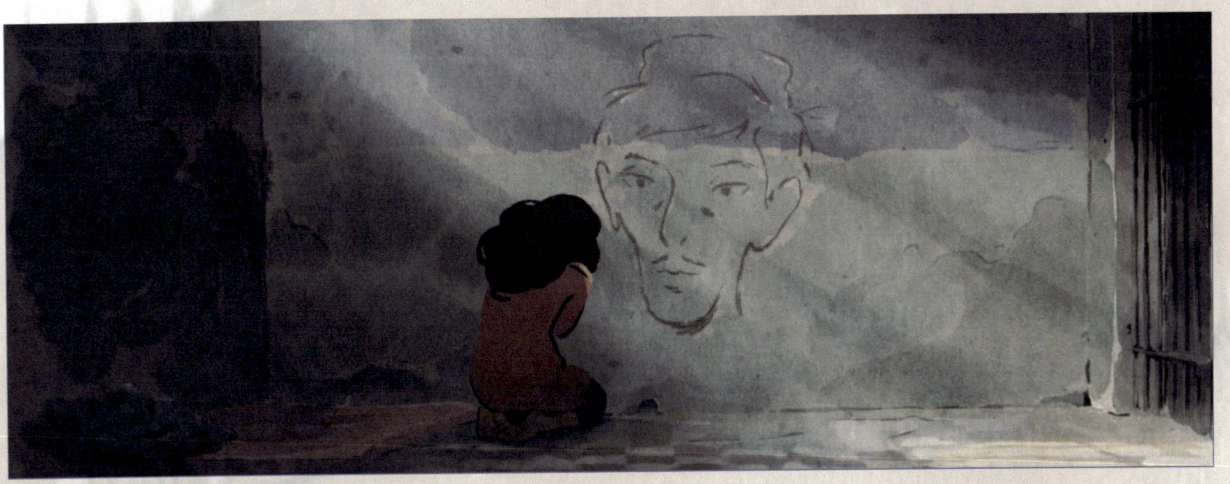

CONCLUSION

Conclusion

Animation in Afghanistan began in the 1970s with experimental works by individual artists. During this period, the Afghan Film Organization (AFO) was established, which played a crucial role in promoting animation production.

The civil war and Taliban regime led to a decline in animation production. Many animators left the country or stopped working due to restrictions and lack of resources.

After the fall of the Taliban in 2001, animation in Afghanistan experienced a resurgence. The Goethe-Institut's animation project, launched in 2010, provided training and support for Afghan animators. This led to an increase in production, including short films, series, and documentaries. In the years following, Afghan animation was thriving, with a new generation of animators exploring various styles and themes. The use of digital technologies increased, and Afghan animators overcame countless challenges to bring their stories and ideas to life.

However, the animation industry in Afghanistan has faced significant challenges since the Taliban takeover in 2021. The Taliban's strict rules and censorship have limited freedom of expression, making it difficult for animators to produce content that is not in line with the Taliban's ideology. Many Afghan animators and artists have fled the country due to fear of persecution.

Some of the current challenges for the animation industry include: limited support from Government; limited funding, resources and infrastructure; gender restrictions; limited academic place; a smaller pool of animation professionals in the country; and limited platforms for showcasing and promoting work.

There are, however, some opportunities – in particular the many untold folktales and stories of Afghanistan. To realise these opportunities, I would make the following recommendations:

- the Government and other non-governmental organizations should invest in and fund the animation sector
- Animation should be officially recognized as part of the higher education system in Afghanistan
- Both local and international instructors should be recruited for trainings
- Animation workshops should be organized to train individuals interested in this field
- Scholarship and study programs should be offered
- The Animation Department of Afghan Film should be reopened
- The AfghAnimation Film Festival should resume once again.
- Funding should be provided to purchase equipment for the animation department of Afghan Film.
- Afghan Animations should be broadcasted by Afghan TV Channels

The animation industry in Afghanistan faces significant challenges, but there is still hope for its revival and growth in the future. There is much creative talent in the people of Afghanistan, and I hope there will be a promising future for animation in the country.

REFERENCES

References:

(2005, June 17). Retrieved January 15, 2020, from festival@3continents.com: http://www.3continents.com/f3c2004/eng_afghanistan.html

(2005, June 17). Retrieved January 15, 2020, from festival@3continents.com: http://www.3continents.com/f3c2004/eng_afghanistan.html

(2011, Dec 1). Retrieved from Yahoo: http://news.yahoo.com/sesame-street-comes-afghanistan- 102156515.html

(2018, November 22). Retrieved from Muppet Wiki: https://muppet.fandom.com/wiki/Baghch-e- Simsim

Almy, H. (2020, January 22). FKH Media. (M. Azizi, Interviewer)

Bayan, M. R. (2018). pp. 37-44.

Bayan, M. R. (2018). The Place of Animation in Afghanistan's Art. Kabul.

Beckerman, H. (2003). Animation the whole story. Newark: Allworth press.

Bendazzi, G. (2003). Cartoons one hundred years of cinema animation. Indiana university press.

Blair, P. (1994). Cartoon Animation . Walter Foster .

Boyd, N. (2019, December 4). Retrieved from study.com: https://study.com/academy/lesson/historical-research-design-definition-advantages-limitations.html

Britannica, T. E. (2016, march 30). Retrieved from Encyclopædia Britannica : https://www.britannica.com/biography/Behzad

Dupree, N. H. (2020, march 2). Retrieved march 5, 2020, from Encyclopædia Britannica: https://www.britannica.com/place/Afghanistan

Dupree, N. H. (2020, march 2). Retrieved march 5, 2020, from Encyclopædia Britannica: https://www.britannica.com/place/Afghanistan

Feroz, N. (2016, feb 3). Retrieved from BBC: https://www.bbc.co.uk/blogs/collegeofjournalism/entries/cc23ed0d-2bea-4043-9a19- 79523e08143a

George, M. i. (2002, January 29). Retrieved december 14, 2019, from BBC News: http://news.bbc.co.uk/1/hi/world/south_asia/1787079.stm

Haqiqi, L. (2019, December). First Female Animator. (M. Azizi, Interviewer)

Hashimi, M. N. (2019, December 27). Afghan Animation . (M. Azizi, Interviewer)

Hedayati, M. D. (2020, March). Animator. (M. Azizi, Interviewer)

Hossaini, M. (2019, December). Animator . (M. Azizi, Interviewer)

Jahanyaar, H. L. (2020, Jan 25). Kabul University Library . (M. Azizi, Interviewer)

Johnson, F. T. (n.d.). The Illusion of Life . Abbeville Press .

Karimi, M. (2020, Jan 13). Zawia Media. (M. Azizi, Interviewer)

Karimi, S. (2019, December 29). Afghan Animation. (M. Azizi, Interviewer)

Kiyan, A. M. (2020, Jan). Mehrgan International Film Festival. (M. Azizi, Interviewer)

Kumar, R. (2018, March 3). Retrieved from npr: https://www.npr.org/sections/goatsandsoda/2018/03/03/590215035/if-afghanistan-ran-the- oscars-the-breadwinner-would-triumph

Lovaan, R. (2020, Jan). Herat International Women's Film Festival . (M. Azizi, Interviewer)

Milligan, M. (2016, May 19). Retrieved from Animation Magazine : https://www.animationmagazine.net/features/internation-al-co-pro-the-breadwinner-enters-production/

Moffitt, K. (2019, November 20). Retrieved from study.com: https://study.com/academy/-lesson/research-methodology-approaches-techniques-quiz.html

Nabipoor, S. (2020, Jan 25). Fine Arts Faculty. (M. Azizi, Interviewer)

Sediqi, H. (2019, December). National Radio Television of Afghanistan Archive . (M. Azizi, Interviewer)

sesame workshop. (2018, December 26). Retrieved from https://www.sesamework-shop.org/what-we- do/shows/baghch-e-simsim

Shahrani, E. a. (1392). Soratgaran Ma'asir Afghanistan, معاصر صورتگران. Kabul: کتب انتشارات بیهقی،Kotob Baihaqi .

Wolfe, J. (2014, March 3). Retrieved from Animation World Network: https://ww-w.awn.com/news/cartoon-saloon-aircraft-productions-co-produce-breadwinner

Made in the USA
Las Vegas, NV
03 November 2026

57424070R00050